CLIMATE CHANGE FOR YOUNG PEOPLE

CLIMATE CHANGE FOR YOUNG PEOPLE

THE ANTIDOTE TO ECO-ANXIETY

DAVID STARK

Copyright © 2023 David Stark

The moral right of the author has been asserted.

Apart from any fair dealing for the purposes of research or private study, or criticism or review, as permitted under the Copyright, Designs and Patents Act 1988, this publication may only be reproduced, stored or transmitted, in any form or by any means, with the prior permission in writing of the publishers, or in the case of reprographic reproduction in accordance with the terms of licences issued by the Copyright Licensing Agency. Enquiries concerning reproduction outside those terms should be sent to the publishers.

Matador
Unit E2 Airfield Business Park,
Harrison Road, Market Harborough,
Leicestershire. LE16 7UL
Tel: 0116 279 2299
Email: books@troubador.co.uk
Web: www.troubador.co.uk/matador
Twitter: @matadorbooks

ISBN 978 1803135 311

British Library Cataloguing in Publication Data.
A catalogue record for this book is available from the British Library.

Printed and bound by CPI Group (UK) Ltd, Croydon, CR0 4YY
Typeset in 11pt Adobe Caslon Pro by Troubador Publishing Ltd, Leicester, UK

Matador is an imprint of Troubador Publishing Ltd

MIX
Paper | Supporting
responsible forestry
FSC® C013604

Advice to my grandchildren:
You need to be a specialist and a generalist. Enough of a specialist not to be fooled by people who think they are clever "know-it-alls" but are not. A generalist to embrace everything that this wonderful world has to offer.

David Stark

FRONTISPIECE

I reckon I have produced the first comprehensive guide to climate change that people in the street can understand, especially young people who have been told by climate activists that their parents' generation has ruined their future. This will start an informed conversation which should have happened a long time ago. Why must we give up our petrol, diesel and hybrid cars? Why must we get rid of our gas-fired boilers and gas cooking? How much of our treasured modern Western lifestyle will we have to abandon to 'save the planet'? Will poorer countries get the message and restrict the progress of their societies, or are there alternative and affordable technologies that can be substituted for the fossil fuels that made Western countries wealthy? Are we actually saving the planet or harming it further with 'renewables' like biofuels which involve cutting down huge areas of forest? How much will the transition cost, how fast will it happen, and how will we know if our sacrifices have been worth it? What government policies are merely costly greenwash and virtue-signalling? Why has Europe effectively been funding Putin's murder of Ukrainians by buying energy from Russia? Politicians had better read this book and come up with the answers.

I believe that the reason no one has so far produced a popular book on climate change is that to do so exposes the uncertainties in climatology and its related sciences in what is a very complex subject. Those who believe that human emissions are having a disastrous effect on the planet, and they mostly believe this honestly and passionately, wish to present a simple message to avoid any doubt. During the period that human carbon dioxide emissions have risen, especially since the middle of the 20th century, the world has warmed. The assumption is made that further emissions will accelerate this warming and the climate will change in ways that can only be bad for the environment and, ultimately, human beings. This book explores such contentions in some detail but also explores alternative approaches, which focus more on the mechanics of natural climate change, as observed throughout history, and which argue that human emissions may play only a limited part in climate change. Warming also has benefits as well as downsides.

If this is true, it might lead to a change in emphasis from political solutions that focus on mitigation measures (ones designed to move the climate back to where it was by radically changing society to reduce human CO_2 emissions) towards focusing more on adapting human society to climate change as it happens. Each annual COP climate conference shows that politicians are good at declaring 'climate emergencies', but poor at actually agreeing and implementing policies which will inevitably place a severe financial burden on their citizens. As I demonstrate, current mitigation measures based on 'renewables' are ineffective (they are poor at reducing net emissions) and do not come out well in cost benefit analyses, but are religiously pursued. It seems that a CO_2VID irrationality virus has infected the Western world while countries like India and China carry on regardless, immune to the contagion, their priority being to free their citizens from poverty.

In contrast, investment in climate/weather adaptation measures has proved very successful and financially profitable: strengthening sea and river defences against floods, warning threatened populations about impending extreme weather events so they can prepare better, increasing search and rescue services for when disasters happen, building stronger buildings and infrastructure that can withstand fierce storms, developing irrigation schemes that protect us against droughts, etc. As the International Disaster Database shows, human deaths from natural disasters are a small fraction of what they were half a century ago. Adaptation has been proved to work and be highly cost effective.

This book is aimed at the general reader, but in particular teenagers and young people at college and university who are thirsting for information on what they have been told is the biggest issue that their generation will face. They need to know the facts. We must question everything if we are to understand this world and where we are going in it. Blind acceptance of the current apocalyptic message on climate change is not healthy. It is scaring our children, almost certainly unnecessarily, because humanity has shown many times before that it can rise to the challenge. I am sure it will do so again with climate change.

David Stark
October 2022

Note: This book was mostly written in 2020 and early 2021, so preceded the late 2021 escalation in energy prices and Putin's 2022 invasion of Ukraine, both of which were predictable.

CONTENTS

Introduction — xiii

1. Scary stuff – or is it? — 1
   Psychology lesson

2. My fears and tears: Nazis, Communists and the end of the world — 23
   History lesson

3. What does the United Nations say about climate change? — 41
   Meteorology and futurology lessons

4. Tipping points and cliff edges — 59
   Futurology lesson

5. Another panic: ocean acidification — 70
   Ecology and chemistry lessons

6. What about species extinction? — 86
   Ecology lesson

7. Why poverty reduction is a more urgent problem than climate change — 104
   Sociology and economics lessons

8.	The chemistry of "greenhouse" gases (GHGs) Chemistry lesson	125
9.	It's the sun, stupid Meteorology lesson	139
10.	What can we do to reduce human emissions? Technology and economics lessons	155
11.	The Garden of Eden complex Psychology lesson	193
12.	Jolly hockey-sticks and cycles within cycles Politics, mathematical graphs and history lessons	221
13.	My conclusions: the skeptics of sceptics challenge Topics for debate	255
14.	Addressing climate change, the storm in a teacup A lesson in pragmatism	267
	Appendix Review of UN IPCC Assessment Report 6	291
	Postscript	329

INTRODUCTION

To a teenager, the prospects for climate change are very scary. The future itself is scary because we can't predict it. You, the young reader, have hopefully been largely protected from the problems of the world by your parents, but you are starting to think about how you will face the future as a more independent human being. You should be very optimistic because there are lots of positive things happening everywhere; but there are also problems that your generation will have to solve.

My generation had its fair share of successes and failures in trying to sort the world out. I have two granddaughters who are asking me more and more questions about life. They are ten and eight, and they are much brighter than I was at that age. They have access to much more knowledge than I did, but they need help to work out what is real and what is imaginary, because the human imagination is incredibly inventive. I try to do this without spoiling the fantasies and dreams that are part of a happy childhood, and which never completely leave us as we grow up.

But there comes a point when we have responsibility for other people and make decisions that will affect them. We must be sure of what is true and what is not. We must be adaptable enough to listen to the knowledge that other people have gathered and the views they hold, and compare these with ideas we already have. The value of a scientific approach to life is that we can find out what facts are correct, and if necessary change our minds to seek good solutions to problems. If we do not learn, we are sure to repeat our mistakes.

The last thing we should do is panic. When this happens we look for quick solutions which may end up making the problem even worse. I believe that this is what is happening today with climate change. There is an issue that we have to face; but there are different solutions, some of which we are not yet properly exploring. I remain optimistic because I have witnessed many problems that we have come to terms with over my last seven decades on the planet. This does not guarantee that we will solve this one, but using words like 'climate chaos' and 'climate emergency' just raises tensions and are of no help. Addressing climate change will be a marathon and not a sprint, and will most likely involve many small changes over many decades, not radical change immediately as some foolish people demand. Revolutions do not have a good track record.

Amazingly, despite this subject being of such widespread interest, some say the most important issue facing us all, there has so far not been a single book published which covers the subject in a way that allows teenagers and 20-somethings to discuss the issue, knowledgeably and constructively. Hence the apparent panic, which does not help those who suffer from the 21st-century ailment, eco-anxiety. Some people say that the issues are far too complex for young people to understand, but I passionately believe that this is not so, and this was the inspiration for this book. When I was a teenager, the technologies involved in computers, colour televisions and mobile phones were new and too difficult for most people to comprehend, or like the internet were not even dreamed about, but are now commonplace and within the grasp of small children. I have every confidence in young people understanding what is at stake.

In this book I tell a number of stories, some of which inevitably stray into broader areas of science, history, philosophy, geography and politics, but please stick with it because it will all make sense – I hope. Climate science is very complicated, as are the ideas people have to change our lifestyles to address the problems we see. I am sorry that this book has to be so long, but it could easily have been longer. Dip into it for reference when you see an aspect of the subject in the news.

I hope you have fun reading this and that you learn a lot. Most of all, I hope it raises lots of questions you will discuss with others, and make life very difficult for your teachers and your lecturers at college, who will be forced to keep up with your level of new knowledge. Do not be afraid to question everything and seek out others with different viewpoints to arrive at the truth.

The world is a wonderful place. People like me have made our contribution, for good and for bad. Only others can judge, if indeed judgement is necessary. It's over to you now. Good luck.

Reader feedback on issues raised by this book can be submitted via the website: www.davidstarkauthor.co.uk.

CHAPTER 1
SCARY STUFF – OR IS IT?
PSYCHOLOGY LESSON

Below are some alarming statements by organisations and people, shown in italics. My less alarmist, and I believe more pragmatic and common sense comments, are after each one.

Activist group Extinction Rebellion: "We are facing an unprecedented global emergency. Life on Earth is in crisis: scientists agree we have entered a period of abrupt climate breakdown, and we are in the midst of a mass extinction of our own making… Human activity is causing irreparable harm to the life on this world. A mass extinction event is underway. Many current life forms could be annihilated or at least committed to extinction by the end of this century. The air we breathe, the earth we plant in, the food we eat, and the beauty and diversity of nature that nourishes our psychological well-being, are all being corrupted and compromised by the political and economic systems that promote and support our modern, consumer-focused lifestyles."

This statement infers that human greed (consumerism) has led to climate breakdown (there is no definition of what "breakdown" means) which is causing a mass extinction event (presumably similar in scale to the event which caused the extinction of the dinosaurs). Politics and economics are the drivers of this problem, in other words, the modern industrial and commercial systems which support our rich, safe, comfortable modern lifestyles. XR is just one of many organisations that uses climate change to justify its political position. The "emergency" is "unprecedented". Like most of the other statements below this is an example of hyperbole, which the dictionary defines as "a figure of speech using absurd exaggeration". XR is worried about "psychological well-being", but eco-anxiety results from such absurd exaggeration, which is the common tool of activist groups. "Scientists agree", but which ones, and how can we check?

This book explores the scientific evidence behind how the climate changes to consider what part humans might play. But statements like the one above make it clear that activists have a political agenda which I will show colours their attitude to evidence that is presented. They seek only information which supports their cause with an inclination to ignore all other data. If humans can be blamed for all of the problems facing the environment, and are largely or wholly responsible for the changing climate, the activists can argue that leaders need to impose drastic changes to obtain their perfect view of how the world should be: their utopia. To achieve dramatic political change, a dramatic and all-pervading problem must be defined.

In Chapter 2 I will begin to explore how political environmentalism has developed during my lifetime, before dealing with mostly scientific matters up to Chapter 11, where I will again consider the psychology and politics of environmentalism. Chapter 12 shows how a deeply held political ideology can blur objectivity, making it very difficult to separate the science from the politics. When activists talk about evil and greedy humans, they adopt the moral high ground with a religious certainty. They tell others how to behave while benefitting from all the systems they say are corrupt. It is funny how most members of XR seem to be white, privileged people living in rich, developed, democratic countries. Thirty thousand climate activists, politicians, bureaucrats, academic researchers and the media fly into an international COP conference on climate change each year, achieve very little, but preach to us that we should fly less to save on CO_2 emissions. This is gross hypocrisy.

Activist group Greenpeace: "Our climate is breaking down. Rising seas and extreme weather events are costing lives and putting tens of millions of people around the world at risk. And younger generations are being robbed of their future on a healthy, liveable planet… The frequency and strength of storms is increasing, leaving destruction in their wake. And rainfall patterns are shifting, causing devastating droughts and floods. As our climate breaks down, billions of people are already struggling to cope and it's the poorest who are being hit hardest… In drier, hotter conditions, wildfires rage out of control, reducing mighty forests to ash. The oceans are warming and the water is becoming more acidic, causing mass coral die-offs and the loss of breeding grounds for sea creatures. Delicate ecosystems that are home to insects, plants and animals struggle to adapt quickly enough to the changing climate, putting one million species at risk of extinction. That means our food security, health and quality of life are all under threat."

Again, we do not have a definition of "climate breakdown". Our children will not live on a "liveable" planet? Is it responsible to tell our young people that they are likely to die? What is the evidence for this? Which human being has so far ever been able to predict the future? It is absurd to say that rising seas are "costing lives" since the UN says that average sea level rise over the last century has been less than 300 mm. People would have to move very slowly to fail to avoid that. The International Union for the Conservation of Nature (IUCN) had about 24,000 species on its endangered list in 2016, so it is not clear how Greenpeace thinks that "one million species" are at risk.

The data does not show that the frequency and intensity of storms are increasing. In terms of deaths from extreme weather events, the International Disaster Database in 2014 noted that in the decade from 2004 to 2013, worldwide climate-related deaths (including droughts, floods, extreme temperatures, wildfires and storms) plummeted to a level 88.6% below that of the peak decade, 1930 to 1939. The year 2013, with 29,404 reported deaths, had 99.4% fewer climate-related deaths than the historic record of 1932, which had 5,073,283 reported deaths from the same category. We are much better now than ever before at protecting people from extreme weather events. In responding to this, Alexander J Epstein (*The Moral Case for*

Fossil Fuels 2014) said: "We don't take a safe climate and make it dangerous; we take a dangerous climate and make it safe." I will say more about this later, and Chapter 5 is devoted to 'ocean acidification'. In response to President Joe Biden's claim that climate change is the most serious problem facing the United States, Michael Shellenberger pointed out that in 2020 nearly 90,000 Americans died from drug abuse compared to only 308 from extreme weather and other natural disasters.

Activist group Friends of the Earth: "The UN Refugee Agency says that climate change adds to many of today's conflicts – from Darfur to Somalia to Iraq and Syria. The effects will get worse as temperatures rise further. So we must cut greenhouse gas emissions to stop temperatures rising. People also need protection from the climate change we can no longer avoid. That means financial support for developing countries and legal protection for climate refugees."

The assumption is made that rising temperatures are causing more droughts, and some organisations have claimed that it is droughts which have caused the conflicts, and not political disagreements. There is no evidence to support this. The UN report in 2013 said that droughts were not increasing globally, and predictions about climate refugees have been incorrect. In 2005 the UN cited a study which said that there would be 50 million climate refugees around the Pacific by 2010 because of sea level rise. They did not turn up as the sea level continued to rise at a modest 2–3 mm per year and, as we will see later, so did the area of most islands due to the seas piling up naturally eroded corals onto beaches.

There is also an assumption that all the warming over the last 200 years has been caused by human emissions, and that if we cut emissions, the warming will stop. As I will show later, at least some of the warming has been due to natural variations in climate. Also, if we deny poor countries cheap fossil fuels, which got rich countries out of extreme poverty over the last two centuries, will we consign them to continued poverty and wars?

Activist group WWF: "Climate change is the greatest environmental challenge the world has ever faced, but we can do something about it. We are the last generation that can stop devastating climate change. We have the knowledge and the tools – we just need politicians to lead the way."

This seems to assert that all we have to do is reduce human greenhouse gas emissions and the climate will return to 'normal', whatever normal is, and that we know how to reduce emissions back to where they were before the industrial era. But this is not a simple problem to solve, at the wave of a politician's magic wand.

A leader of activist group Population Matters and naturalist David Attenborough: "Right now we are facing our greatest threat in thousands of years, climate change… Our climate is changing because of one simple fact. Our world is getting hotter… Scientists across the globe are in no doubt that, at the current rate of warming, we risk a devastating future… The science is now clear that urgent action is needed."

As we will find out later, the climate has only been getting warmer (a modest 1°C) for the last 150 years from a very cold period called the Little Ice Age, when the River Thames sometimes froze over in winter. It would be more accurate to say that our world is getting less cold, or warmer, not "hotter". Scientists are not in agreement that we face a "devastating" future, just that human emissions are contributing to the warming. As I will show later, the science is far from clear, which makes our response to the issue difficult to agree on.

Peter Stott (the UK Met Office and the University of Exeter): "What we've seen is this steady and unremitting temperature trend. Twenty of the warmest years on record have all occurred in the last 22 years… That warming trend cannot be explained by natural factors, but is caused by human activities, in particular by use of fossil fuels."

As we will see, the temperature trend has not been "steady and unremitting", with two periods of warming in the 20th century (1920–45 and 1975–98) and a period of cooling (1945–75), leading to a headline in *The Guardian* newspaper which foolishly said on 29 January 1974: "Space satellites show new Ice Age coming fast." The Nimbus satellite programme, which started in 1964, showed rapidly increasing Arctic ice. Again, the contention is made that only human activities are causing the warming, but there are many other scientists who are studying the 'natural' factors that contribute to it.

Activist Greta Thunberg in America in 2019: "Why should we go to school when there is no future? This is all wrong. I shouldn't be up here. I should be back in school on the other side of the ocean. Yet you all come to us young people for hope. How dare you! You have stolen my dreams and my childhood with your empty words. Yet I am one of the lucky ones. People are suffering. People are dying. Entire ecosystems are collapsing. We are at the beginning of a mass extinction, and all you talk about is money and fairy tales of eternal economic growth. How dare you."

Again, no evidence is presented about the collapse of ecosystems and of mass extinction, and the assumption is made that economic growth is at the heart of the problem. I will consider later if this has any merit, or if the inevitable poverty caused by deindustrialisation (the proposed solution) would make matters worse. Child soldiers and teenage eco-warriors obtain their ideology and weapons from adults, and Kevin Anderson (University of Manchester and Uppsala University in Sweden) is one of Greta's advisors. He said to a group of young students (the lecture is on Youtube): "We are all going to hell in a handcart at the moment. We're all going somewhere we don't want to be. There are virtually no positive stories out there at any large-scale collective level… Every day we choose to fail [and] now we are passing on generationally the burden to you here… The taboo issue of the huge asymmetric distribution of wealth underpins the international community's failure to seriously tackle climate change." He says we need to "disturb the dominant socio-economic paradigm of ongoing growth, with resources, power and CO_2 skewed to a privileged few." His extreme left-wing politics are obvious.

Given the horror stories told by prominent activists, it is understandable that young people are scared about their future, especially particularly vulnerable ones like Greta Thunberg with Asperger Syndrome (AS). Anxiety and mood swings are the symptoms. In an interview she revealed that she had become concerned about the environment when she was seven or eight (about the time that David Attenborough and Al Gore were scaring children about the imminent fate of polar bears, which we have now found out was unwarranted). At 12 she couldn't understand why her parents and her classmates didn't seem to care about the issues. "Everything is wrong. Everything is so strange. Everything is so sad, and why isn't anyone doing anything about this? And so I fell into a depression. It lasted for maybe a year. I stopped talking. I only spoke to some people; my teacher, some members of my family, and I stopped eating almost entirely. I lost a lot of weight. I was so depressed. Nothing seems to matter anymore".

Her mother, Malena Ernman, writing in *The Observer* in February 2020, described the difficulties she had bringing her daughter up. "Greta was 11 and was not doing well. She cried on her way to school. She cried in her classes and during her breaks, and the teachers called home almost every day." Describing her first panic attack: "She makes a sound we've never heard before, ever. She lets out an abysmal howl that lasts for over 40 minutes." She changed when her campaign got going: "Greta's energy is exploding. There doesn't seem to be any outer limit, and even if we try to hold her back she just keeps going." Her mother is obviously happy that her condition is better, but when the limelight fades and reality beckons, one wonders what the outcome will be. Greta's AS no doubt contributed to the malaise, and she herself suspected "cognitive dissonance" from being on the autism spectrum, but she regards her condition as a "super power", allowing her to be super-focused. Perhaps she is just super-obsessed. Her childlike frailty and ostensibly noble intentions protect her from critical questioning by those media people she is allowed to be interviewed by, so it is difficult to know which facts she bases her apocalyptic world view on.

There are always two sides to any story. Climate activists focus on the bad news, and then exaggerate it a bit. They will look at the warm summer of 2003 and point out that many people in Europe died during the heatwave. They will ignore the 2015 *Lancet* study of 74 million deaths between 1985 and 2012 in 13 countries which found that 7% were associated with low temperatures and 0.4% with high temperatures. Warmer temperatures are better for us. In the middle of the Little Ice Age, during the 17th century, a third of Europeans died from cold, malnutrition, disease (because they were weakened by poor nutrition) and conflict due to scarce resources. The world's temperature is now only about 1°C warmer than during the Little Ice Age. There are benefits from having a slightly warmer climate. Most people in the world would prefer longer summers and shorter winters.

I have also made the point above about our ability to adapt to the climate and deal with extreme weather events, whether they have been caused by anthropogenic global warming (AGW) or not. A study by Delft University of Technology in April 2018 in *Nature Communications* found that flood deaths in Europe have been falling by about 5% per year for the last 60 years, with financial losses reducing by 2% per year. It concluded that a warming climate was actually helping because there was less risk from sudden thaws. It reviewed 1,564 damaging floods between 1870 and 2016 in 37 European countries and "extreme hydrological events went down slightly in the 21st century". It concluded that there were more people in urban areas, houses and flood defences were stronger, and it was easier to evacuate people by helicopter.

AGW should be addressed, but panic is not a good response. As Hans Rosling said in his book *Factfulness*: "Fear plus urgency make for stupid, drastic decisions with unpredictable side effects. Climate change is too important for that. It needs systematic analysis, thought-through decisions, incremental actions, and careful evaluation. Exaggeration undermines the credibility of well-founded data: in this case data showing that climate change is real, that it is largely caused by greenhouse gases from human activities such as burning fossil fuels, and that taking swift and broad action now would be cheaper than waiting until costly and unacceptable climate change happened. Exaggeration, once discovered, makes people tune out altogether."

> **Discussion: Why do bad news stories make the headlines much more often than good news stories? Why are we usually more conscious of threats than opportunities? Can over-exaggerating problems make people switch off and take no action at all?**

This book seeks to give a balanced analysis of climate change, and since the above quotes illustrate how a large amount of alarming and inaccurate information leads to a very pessimistic outlook, I will be accused of being too optimistic. In trying to obtain a balanced perspective, and concentrate on facts rather than hyperbole, I seek to give young people sufficient information to discuss the subject constructively, rather than just accept what politicians and activists insist on. Most of all I seek hope and not hype.

When we are considering the climate we must remember that this is not the same as the weather. When we see extreme storms or heatwaves on our televisions, or even droughts that last as long as five years, these are not examples of climate change, or any indication that climate change is happening. They are weather events. The dictionary definition of climate change says that we must see a trend happening for many decades, 30 years being a good guide, although there are some natural climate cycles that only reveal themselves over longer timescales. As the world warms there will inevitably be record temperatures from time to time (at least since records began in the 19th century), but the 1930s is still the decade with the most US temperatures over 100°F, the time of the Dust Bowl in the Midwest, and the last decade has seen the second lowest number of large hurricanes making landfall in America since records began in the 1850s (Figure 6). News reports of extreme weather events are certainly increasing.

We have had satellites looking down on our planet for more than 40 years, and they are probably our best guide to detect global trends over this period. What do they tell us?

Figure 1 shows how much greener the planet became between 1982 and 2009. Information compiled from NASA's Moderate Resolution Imaging Spectrometer and the US National Oceanic and Atmospheric Administration's (NOAA) Advanced Very High Resolution Radiometer measures what is called Leaf Index; in other words, how much leaf cover there is in various parts of the world. While some areas have lost vegetation, most have gained it, by as much as 14%. This is good news. We will consider deforestation in Chapter 7, because there are problems with this in some parts of the world, but generally the world is getting greener, and this is also a good news story for animals like us which depend on vegetation. Many scientists believe that the causes of this are warmer weather, with warmer cyclones able to hold more water, and more CO_2 in the air that is basically plant food.

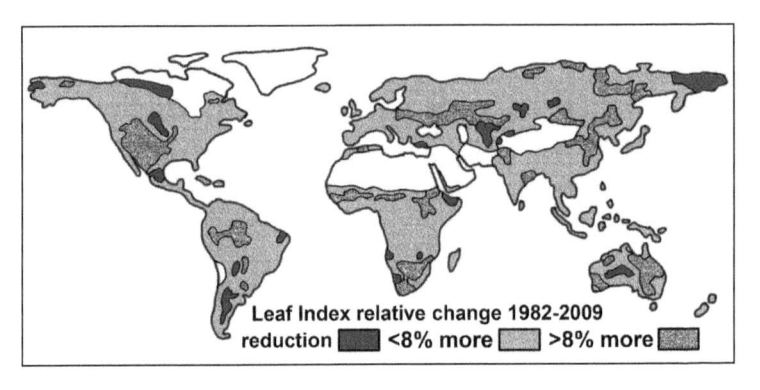

Figure 1: Percentage increase in leaf index 1982–2009.

More CO_2 also makes plants more drought resistant. When a plant opens its pores to ingest CO_2, H_2O escapes so the less it has to do when there is a higher concentration of CO_2 in the atmosphere, the less moisture it loses. Notice how the area just south of the Sahara Desert, the Sahel, has one of the highest levels of greening. And yet climate activists claim that more droughts in this area are causing resource conflicts which terrorist groups like Boko Haram and al-Qaeda exploit. "Military chiefs fear these conflicts could spill over into neighbouring countries and exacerbate the migrant crisis" (*The Times* 10 March 2021), but the newspaper is linking various issues: alleged but erroneous climate change analysis along with terrorism and migration. This is bad journalism based on bad science to justify a false ideology.

It may be the same good news story in the oceans due to warmer seas and more CO_2 in them. According to a study reported in *Science* magazine on 26 November 2015, phytoplankton in the North Atlantic has increased tenfold since 1970, benefitting a wide range of ocean life as food for zooplankton and fish larvae. Phytoplankton account for about half of the photosynthesis and oxygen production on the planet, yet are only about 1% of the global plant biomass.

We often see dramatic pictures of bush fires in California and Australia, natural occurrences in these regions, but allegedly more likely/serious because of global warming. This may or may not be so, but the satellite data (Figure 2, an Africa extract of global data) show that the area of the planet suffering from wildfires is reducing. Whatever is happening with changing climate, people become more resourceful as societies get richer, and they can afford firefighting facilities to protect crops and homes from wildfires which encroach onto their land.

We can learn from indigenous peoples in California and SW Australia who carried out controlled burns in winter to provide fire breaks in summer. Tim Blair in the Australian *Daily Telegraph* at the time of wildfires at the beginning of 2020 wrote: "Natural ecosystems are now, across thousands of hectares of national parks, nothing but cinders and ash. Enjoy your protected habitat, little ground-dwellers. Those woodland creatures would have been better off under the stewardship of my grandfather, whose Aboriginal-style controlled burns were not limited to his own property and its surroundings. Every year he would burn the long grass alongside local roads to make those roads more effective as fire breaks. He never sought permission from the local fire brigade to light these fires because my grandfather was the local fire brigade captain… Brian Williams, captain of Kurrajong Heights fire brigade told a weather station during a brief break in his ninth week of battling a monster blaze north of Sydney, 'We've been burning less than 1% of our bushfire-prone land for the past 20 years. That means every year the fuel load continues to build'."

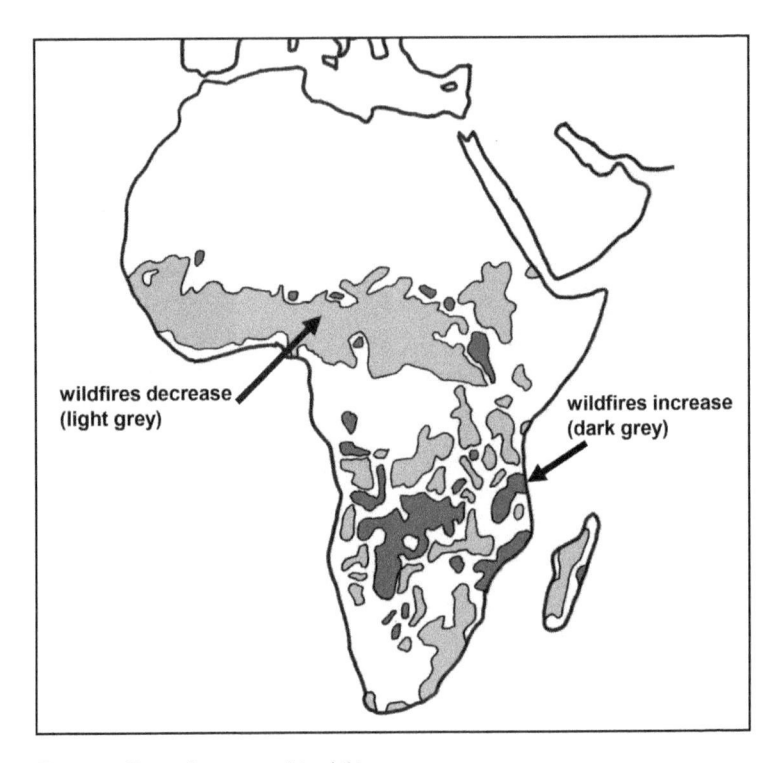

Figure 2: Burned area trend in Africa.

That was why the fires burned for so long. The human failure was not emitting an invisible, uninflammable greenhouse gas, but abandoning traditional land management methods for New Age environmentalism. It is the same in California. ProPublica news agency 28 August 2020: "Between 1982 and 1998, California's agency land managers burned, on average, about 30,000 acres a year. Between 1999 and 2017, that number dropped to an annual 13,000 acres."

Figure 2 is not all good news, because it shows increasing areas of fire in certain parts of south-central Africa, in a band that includes Zimbabwe, Angola, Zambia and Mozambique. Poverty in these places is the most likely factor, with primitive slash-and-burn agricultural techniques still taking place because the farmers cannot afford modern agricultural equipment.

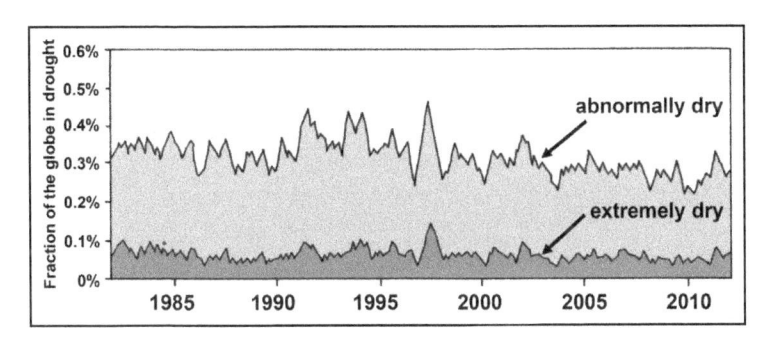

Figure 3: Fraction of the globe in drought from 1982 to 2012.

Figure 3 was taken from an article by *Hao et al.* in *Nature* in 2014, and it shows no trend in the extent of droughts in the period from 1982 to 2012 in various parts of the world. If anything, there has been a slight downwards trend since 1998. A warming climate should generally mean that cyclones hold more rain, but changes to the climate will have different effects in different areas, with droughts in some regions at some times. However, it seems that warming temperatures do not inevitably lead to widespread droughts.

Activists intuitively assume that more heat in the global weather system will produce more storms and they will be more extreme, but this is not borne out by the data in Figures 4 and 5. Global warming does not increase the overall energy in the system, although if warmer cyclones hold more water vapour, this should lead to more flash floods. Figure 5 seems to indicate no trend in major hurricanes (greater than 96 knots) in the last 75 years, and the trend of all hurricanes (greater than 64 knots) is slightly downward. There is a contention that global warming should be more pronounced at the poles than the equator, which would mean that the temperature gradient between the equator and the North Pole would be smaller and storms would be less intense. This is consistent with the data in Figure 5 although, without further evidence, I regard this theory as conjecture.

The southern hemisphere is less subject to climate changes because of the huge buffering power of the extensive southern oceans and the enormous amount of Antarctic ice, which is why we tend to focus our attention on northern hemisphere climate phenomena.

The UN *IPCC Assessment Report 5* (2013) said that there is "low confidence" in the long-term increase in intensity of tropical cyclones, its way of saying that it is highly unlikely. If

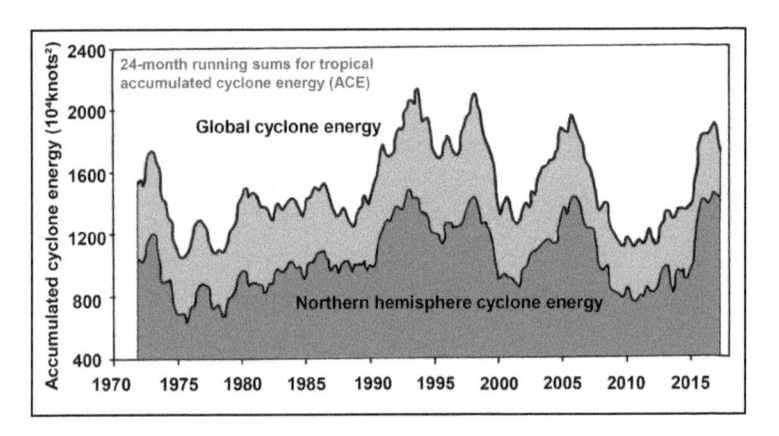

Figure 4: Global and Northern Hemisphere Tropical Accumulated Cyclone Energy (ACE). Source: Ryan N Maue of NOAA, previously Cato Institute.

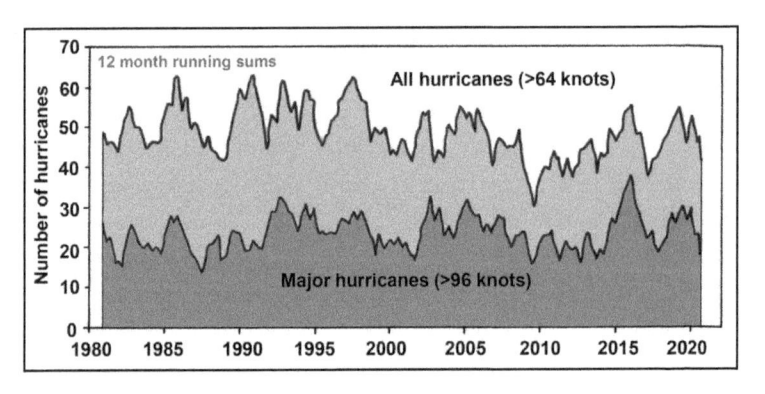

Figure 5: North Atlantic and Western Pacific hurricane landfalls (about 70% of global hurricane landfalls). Source: *Weinkle et al.* 2012, and *Roger Pielke Jr* 2022.

you take a particular time series (a period of time on a graph) in Figure 4, say 1979 to 1994, the intensity of cyclones appears to be increasing, but if you select another one, say 1998 to 2012, it is decreasing. The constant variability of (natural) climate change makes it extremely difficult to identify longer-term trends.

Figure 6 indicates that there is no upwards trend in strong hurricanes (Category 3 and over) making landfall in the USA, where we have good records going back to the 1850s. There does not seem to be an explanation for the seemingly random variation in hurricane activity over time. In the decade after Hurricane Katrina in August 2005, there was the longest lull in major hurricanes on record, but we tend to remember the devastation that this event caused to New Orleans, with over 1,200 deaths after levees burst.

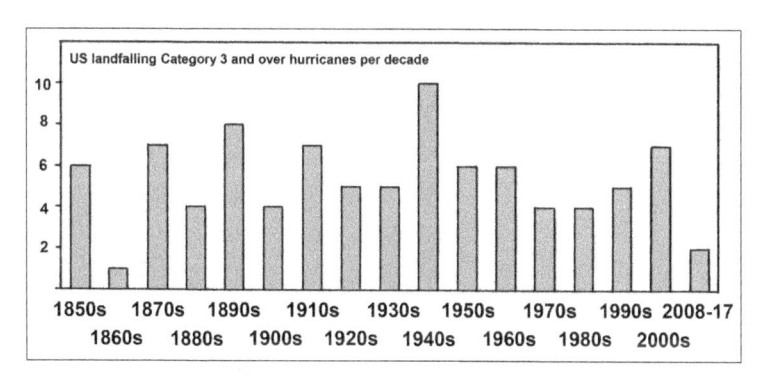

Figure 6: Number of US hurricanes of Category 3 or over making landfall by decade.

There are various points I wish to draw the reader's attention to in this chapter:

1. Activists take what the UN IPCC says and exaggerate/corrupt it to suit their biased, political objectives.
2. The UN is probably happy with this hyperbole if it helps to force governments take action on climate change.

3. The UN does not have a spokesperson who can correct the misconceptions of activists and moderate the alarmism, or indeed answer criticisms people have about the IPCC's own work.
4. Therefore, when an organisation like the BBC wishes to interview someone about a particular issue on climate change, they usually turn to activist organisations like Greenpeace for comment.
5. Activists often use individual extreme weather events to 'prove' climate change and are immune to alternative explanations, for example, 2019/20 Australian wildfires which WWF claimed killed and displaced "almost three billion animals", where poor environmental management was a key factor.
6. The media deals with news, and individual extreme weather events, the more dramatic the better, make news, not the long-term trends that indicate that such events are not becoming more frequent.
7. There is therefore no unbiased and dispassionate point of reference for anyone wishing to seek information on climate change and the political policies arising from it.
8. All of this means that informed debate does not take place, and robust development of practical policies through a democratic process is not possible.

A good illustration of the difference between weather and climate is contained in records of the level of the River Nile at Roda near Cairo, which go back to the year 622 and which effectively record rainfall over 10% of the African continent. In individual years during any era the level can vary from less than one metre to over five metres (the weather). The 30-year trend (the climate) shows a low point just over one metre in the 8th century (during the European Dark Ages or the Early Medieval Period as it is now referred to) and a high point around four metres in the 12th century (the Medieval Warm Period – see Chapter 12 for an analysis of cool/dry and warm/wet cycles over the last 1,500 years). A six-metre year could be in the middle of a period with an average of only one metre.

When I have heard an activist make bold and alarming statements, I have asked for the data, but it has never come. I might be referred to a website which is supposed to provide useful information, but it is very limited and certainly not balanced. What is most worrying is that activists quickly take up an aggressive stance and accuse me of being a 'denier' of science, even though they cannot point me to the relevant scientific evidence. They want me to choose which side I am on – the moral high ground or the greedy exploitation of the planet – when I am seeking a common understanding through rational discussion. They often claim that "the debate is over". Proper debate has never happened because the public was convinced that we should leave the science to the academic researchers who advise politicians. We rely on short news reports on individual research projects, with dramatic headlines grabbing our attention, or reports from organisations like the UN IPCC, which itself never publishes a summary of its findings that people in the street can buy and understand. That is not good enough. This book seeks to fill the information gap.

I have approached individual scientists to try to find answers. Calling someone a 'climate scientist' can be a misleading misnomer. There are some people who study certain aspects of the weather, some people who study polar ice, or geology, or the Sun, or the oceans, or the stratosphere (upper atmosphere), or the troposphere (lower atmosphere), or various fields of biology, or tree rings, or mud sediments, or volcanoes, or cosmic rays, or statistics, or computer modelling or a host of other aspects that have a bearing on climate research. They are very candid about the uncertainties in their specialist subject – if there were not uncertainties they would not obtain research funding – but seem convinced that the overall alarm about climate change is justified. And yet, when I ask them about something outside their area of specialism, they say they are not qualified to comment. I quickly find that I am more of a 'generalist expert' than they are, because I have studied for a long time the many elements feeding into what is casually referred to as 'climate science'. Individual scientists seem no better qualified than me to come to general conclusions.

When we become emotional about something troubling us, our instinct is to differentiate the right from the wrong, the good from the bad, the goody from the baddy. This was expeditious when faced by an imminent threat on the African plains, but it does not help in the modern world to distinguish the true from the false, which must be done with a completely non-judgemental, neutral mind. A scientific approach cannot be moralistic because it must have complete impartiality. Once we know the truth we can work out how to apply it for our benefit and those around us. A prejudiced activist may only notice data which reinforces the prejudice, and be immune to reality. This is not just important for members of the public to appreciate in trying to sort out the science from the political rhetoric but, I will consider later, whether key establishment figures, like James Hansen, director of the US NASA Goddard Institute from 1981 to 2013, and Kevin Anderson, former director of the UK Tyndall Centre for Climate Change Research, were impassioned scientists or passionate activists, because I contend that they cannot be both.

Climate alarmists brand those who support their thesis as 'goodies' and those who do not as 'baddies'. They contend that we must listen to what the former say and ignore the latter because they 'deny' the truth. I treat both with healthy scepticism, whether they agree with some of my conclusions or not.

Our state of mind will also affect our perception of what is true and false, depending on whether we start off as optimists or pessimists. An optimist might regard a mistake as a useful learning experience in the never-ending challenge of personal and human progress. A pessimist might regard a mistake as evidence of fallibility, or that they are basically a bad person. And you must be a bad person. And the system is bad. Perhaps all of humanity is inherently sinful. This cascade of irrational, depressive behaviour is well known to clinical psychologists, and I believe it presents itself in climate change alarmism as eco-anxiety.

I am an optimist but I have darker thoughts that I struggle to deal with. I have been subject to feminist activism all of my life such that I sometimes feel guilty for being a man. It seems that all of women's problems lie at the feet of nasty males like me. I am reminded that my gender has been relegated to second class status every time a woman walks in front of me to make me change direction, as if I am invisible, although it is just a pragmatic convention for males to give way to females. I sometimes wonder why women bring up male children when they will inevitably turn into nasty, lustful men, who might take away women's voting rights and chain them to the kitchen sink if they got the chance. Fortunately my logical side takes over and I give myself a mental slap to get back to reality. I often think that Extinction Rebellion activists need a good reality slap by experiencing what their lives would be like without fossil fuels, effectively a harsh, pre-industrial subsistence struggle.

News is unhelpful in determining the truth as it contains unrepresentative snatches of mostly bad things. The news is perennially depressing. Good things happen under the radar, and usually slowly. The truth is contained in boring data, collected over as long a period as possible. The news will report an incidence of famine, but will not report that famine has been virtually eradicated over the last half century in all parts of the world with stable government. The news will feature dramatic forest fires threatening homes without reporting that the extent of wildfires has reduced over time as humans become more resourceful and keep them under control.

Even more unhelpful is social media which, since 2009 when the 'like' button was added to Facebook and the 'retweet' button to Twitter, has directed users to like-minded posts and taken away the perspective that allows balanced judgement. Algorithms produce echo chambers and thought bubbles. Within a matter of weeks, the 'truth' is defined by peers, often young opinionated people with little experience of life. Greta Thunberg has 5 million Twitter followers and 13 million on Instagram. This is all dangerous for the mental health of youngsters.

It is important to question the established position on climate change, because the scientific establishment has been very wrong before. One hundred years ago, eugenics was the settled science, where world domination by white northern Europeans was justified by their proven genetic racial superiority. One could not deny it. By producing the Reformation, the Enlightenment and modern science they had demonstrated their inherent superiority. This belief did not work out well for 'inferior' black people in America, those subject to British imperialism, Jews and Slavs under the Nazis, and black Africans under apartheid. I am seriously concerned that there are similar dangers with poorly defined climate 'science' that could harm as many people. Utopian 'net-zero' energy policies cause fuel poverty in developed countries, will hinder the progress of developing countries and may cripple the economies of 'self-righteous' Western democracies, while empowering 'pragmatic' autocracies like Russia and China which refuse to become involved. I can only hope that scientific reality wins over the emotional 'saving the planet' ideology before too much damage is done. The planet has been perfectly capable of saving itself for billions of years. Liberal democrats need to save ourselves first.

Discussion: Who can we trust to give us the facts on climate change? Consider various sources: state-funded broadcasters like the BBC, commercial broadcasters, government agencies like the UK Met Office and NASA, activist organisations like Greenpeace and Friends of the Earth, 'independent' thinktanks like insideclimatechange.org or the Global Warming Policy Foundation, and individual bloggers. We are looking for the truth, the whole truth and nothing but the truth.

I was inspired by Hans Rosling's bestselling book *Factfulness* (if you haven't read this book, let it be your next one). His research highlighted how experts, even in their own fields, had an over-pessimistic view of human progress, often holding perceptions that were decades out of date. On climate change, Hans saw the statistics on melting Arctic ice and believed that there was a problem, but he baulked when he was asked by US vice president Al Gore to manipulate the data towards worst case scenarios because "we need to create fear". Perhaps he was aware of HL Mencken's 1918 observation: "The whole aim of practical politics is to keep the populace alarmed (and hence clamorous to be led to safety) by menacing it with an endless series of hobgoblins, most of them imaginary."

Rosling recognised that, in a climate of fear, human beings adopt illogical responses which are counter-productive and have unpredictable side effects. He also observed that economic and climate systems are exceedingly complex, and the various elements involved interact in such a chaotic fashion that it is impossible to make long-term predictions. And yet many people predict what the climate will be like several decades away, if we do not act immediately on emissions reduction. Activists propose solutions which involve technologies that are often speculative or experimental, and therefore impossible to realistically assess in cost benefit analyses. "Thanks to great satellite images, we can track the North Pole ice cap on a daily basis. This removes any doubt that it is shrinking from year to year at a worrying speed. So we have good indications of the symptoms of global warming. But when I looked for the data to track the cause of the problem – mainly CO_2 emissions – I found surprisingly little… Climate change is way too important a global risk to be ignored or denied, and the vast majority of the world knows that. But it is also way too important to be left to sketchy worst-case scenarios and doomsday prophets." I will explore in more detail the 'symptoms' of global warming, like Arctic sea ice extent, as well as the link between CO_2 emissions and global warming, to see if the data is as poor as Rosling suspected.

Rosling gave sage advice when we get anxious about issues like climate change. "I don't tell you not to worry. I tell you to worry about the right things. I don't tell you to look away from the news or ignore the activists' calls to action. I tell you to ignore the noise, but keep an eye on the big global risks. I don't tell you not to be afraid, I tell you to stay coolheaded and support the global collaborations we need to reduce these risks. Control your urgency instinct. Control all your dramatic instincts. Be less stressed by the imaginary problems of an overdramatic world, and more alert to the real problems and how to solve them."

Hopefully this book will give you my understanding of the facts and direct you away from "imaginary problems in an overdramatic world", having studied the subject for many years. When I wrote this book, mostly in 2020, I had no axe to grind and was not pressured by any authorities to observe political correctness, having retired from business life where I could not afford to be controversial and upset potential customers. I believe I have an unbiased outlook and I can be bluntly honest about my findings. But this is a complicated subject, and I welcome any comments and criticism, unlike all activist groups and even government agencies which promote the political warnings about climate change.

Discussion: Is it fair for developed countries to ask poor ones not to use cheap fossil fuels the way developed countries did in the 19th and 20th centuries? Should rich countries pay for their more expensive renewables since it was rich countries which have historically made the most emissions?

'Green' activists would agree with the dictionary definition: "concerned with the future of the Earth's environment". We all must be. But there are other meanings of the word listed. Consider these as the book progresses and decide which might be applied to those who comment on climate change:

- Inexperienced, naïve
- Not fully processed or treated
- Having a sickly appearance
- Jealous

CHAPTER 2
MY FEARS AND TEARS: NAZIS, COMMUNISTS AND THE END OF THE WORLD
HISTORY LESSON

On Friday, 26 October 1962, I said goodbye to my school friends because we did not think we would see each other again. We were ten years old, and we were all about to die. We did not expect to make it until Monday. It seemed inevitable that a nuclear war was about to start, and our school in Glasgow was only 80 km away from the American and British nuclear submarine base targets on the Holy Loch and Gare Loch. We would be among the first on the planet to be blown to kingdom come, or suffer a painful and prolonged death from radiation sickness. All American, European, Russian and Chinese cities were targeted, along with any place that America, France or Britain had military bases. There was nowhere to hide. We were powerless. We were very frightened. This was Cold War anxiety at its peak.

The USSR was building nuclear missile launch sites on communist Cuba, which was only 140 km away from the USA mainland in Florida. President John F Kennedy could not allow this to happen and took part in a stand-off with his opposite number Nikita Khrushchev.

Fingers were on triggers and we found out later how close it came to them being pressed. One US spy plane was shot down and this could have sparked the conflict. On the Saturday the two leaders came to a deal. It looked like Kennedy had won, but we found out later that part of the agreement was that American bases in Turkey would be dismantled as well. This was the closest that the world has come to Armageddon in my lifetime. Both sides reckoned that a third of the human race would have been killed in a matter of hours if hundreds of nuclear weapons had been launched. The fallout of dust from the mushroom clouds might have been similar to that of the meteor explosion that wiped out the dinosaurs.

The fear of how to survive a nuclear war stayed with me throughout the Cold War. How could I possibly protect my young family in the 1980s? It was only when the Berlin Wall fell in 1989, and the Soviet Union collapsed, that my anxiety level dropped. We can look back and see that the MAD (mutually assured destruction) stand-off, where if any country launched a nuclear attack it would have led to disaster for both sides, actually kept the peace, but at the time everyone was under a cloud of threat and fear.

Since the Hiroshima and Nagasaki bombs dropped on Japan at the end of the Second World War, a new generation of hydrogen bombs had become much more powerful. They could no longer be dropped from planes which would not have been able to escape the blast in time. Rockets with nuclear warheads were the solution, and for America, also trying to send a man to the moon, this was achieved with the help of a German Nazi.

The teenage Werner Von Braun was inspired by science fiction dreams of space travel, and joined a rocket club near Berlin, in which he and his fellow enthusiasts tried not to blow themselves up in a quarry as they experimented with various explosive fuels. In 1932 his talent was noticed by the German army which was developing new weapons. Adolf Hitler came to review progress in 1939, but Von Braun could not control the direction a rocket went in, that was, if it managed to get off the ground without exploding. Funding was stopped until SS leader Heinrich Himmler became interested. He could also supply labour for the venture from concentration camps, with thousands of slaves worked to death in the rocket construction and test facilities. Von Braun became an officer in one of the cruellest regimes in history.

He borrowed ideas from American Robert Goddard for liquid oxygen and alcohol fuel to ensure a smooth propulsion, and a spinning gyroscope to keep the rocket on course. Soon his V2 rockets were raining down on London, but the end of the war stopped this programme before more than a few thousand people were killed. The Americans rushed to catch Von Braun before the Russians so they could use his skills for their own weapons development, ignoring the fact that he was a war criminal. He also invented the Saturn rocket which took Neil Armstrong and Buzz Aldrin to the moon in 1969. He became an American hero, and a research unit at the University of Alabama was named after him.

What can we learn from this? When our security is threatened, or a great political venture is launched, like winning the space race by sending a man to the moon to prove that a democratic country is better than a communist one, moral scruples about using the knowledge of a collaborator in Nazi mass-murder are conveniently set aside.

Discussion: Werner Von Braun argued that he did not have any choice but to collaborate with a nasty regime in Germany, but by doing so he learned enough to help the free world later. Others argued that he ignored the suffering of others to advance his passion for rocket flight. What would you have done in his circumstances? Were the Americans right to take advantage of a war criminal from a defeated enemy to try to defeat another enemy? Was it a coincidence that the atomic bomb, for military purposes, was invented before the atomic power station, for civilian purposes?

NUCLEAR WEAPONS ARE DANGEROUS BUT THIS DOES NOT MEAN NUCLEAR POWER IS TOO

The fear of nuclear weapons led to opposition to nuclear power generation for civilian uses by many environmentalist groups, despite the benefits. Around the time of the 1979 Three Mile Island nuclear power station accident, in which no one died, the film *The China Syndrome* was released, starring Jane Fonda. This was a crazy science fiction film which envisaged a nuclear power station core overheating and the reactor burning its way through the Earth's core to China. Fonda became a campaigner against nuclear power, and every heterosexual man who had seen her take off her spacesuit in the 1968 film *Barbarella* was in love with her. The low level of leaked radiation affecting those living close to Three Mile Island was about the same as people living in the Granite City of Aberdeen would naturally experience in a year. Like fellow actor Leonardo di Caprio, Fonda is now a climate change activist. More recently, in her 1994 book *Nuclear Madness*, Nobel Peace Prize winner Helen Caldicott wrote: "As a physician, I contend that nuclear technology threatens life on our planet with extinction if the present trends continue; the food we eat, the water we drink will soon be contaminated with enough radioactive pollutants to pose a potential hazard far greater than any plague humanity has ever experienced."

We have an irrational fear of nuclear radiation, as if it is some invisible, unnatural, unstoppable evil force that humankind has invented. Despite the drama of the 2019 TV series *Chernobyl*, the World Health Organisation calculated that only 54 people died as a direct result of the radiation leak, and this rose to around 230 when the longer-term incidence of cancer deaths was analysed. No deaths occurred from radiation leaks at the only other Level 7 (the highest) nuclear accident at Fukushima in 2011. Chernobyl (1986) was a nuclear power station which had a steam explosion (it could not have been a nuclear explosion because the fuel was not weapons grade) that released radiation into the local area, but it dispersed quickly. The escape of 25 kg of cesium from Chernobyl had the ability to kill 25 million people in a concentrated form but, on the basis that "dilution is the solution to pollution", it was blown over such a large area that one would have had to collect and consume the deposits from a surface area of 20,000 m^2 to kill one person.

Natural radiation is all around us, in the stone buildings we inhabit, and in the sand we walk on at the beach. We obtain about 50% of our exposure to radiation from radon gas in the ground, 15% in medical procedures, 9.5% in foodstuffs and 12% from ionising radiation due to cosmic rays emanating from galaxies outside our solar system. When we eat a bag of Brazil nuts or have a dental X-ray we are exposed to 0.005 mSv of radiation. A typical hospital X-ray causes 0.02 mSv exposure, a London to Tokyo return flight 0.14 mSv and a whole-body scan 9 mSv, the same as the residents of the town next to Chernobyl after the accident. To kill a breast cancer tumor, a dose of 100 mSv is applied. The original name for an MRI was Nuclear Magnetic Resonance Imaging, but they dropped the first letter of the acronym for marketing purposes.

According to the Paul Scherrer Institute in Switzerland, nuclear power had the best safety record in the energy industry from 1970 to 1992 (including non-radiation deaths from day-to-day health and safety issues). In terms of deaths per trillion kilowatt hours of energy produced, nuclear has killed 90 against coal (global) 170,000, oil 36,000, biofuel/biomass 24,000, coal (US) 15,000, gas 4,000, hydro 1,400, solar 440 and wind 150. More people have been killed falling off roofs when installing solar panels than have died because of nuclear accidents. Nuclear waste, although radioactive, is tiny in comparison to the energy it generates. As Steven Pinker (*Enlightenment Now: The Case for Reason, Science, Humanism and Progress* 2018) points out: "Mining uranium for nuclear energy leaves a far smaller environmental scar than mining coal, oil or gas, and the power plants themselves take up about one five-hundredth of the land needed by wind or solar… It has a lower carbon footprint than solar, hydro and biomass." But most environmentalists continue to oppose it, on an emotional, not logical, basis.

Any new technology has the potential for good and bad. It will provide some benefits but have some downsides. We need to maximise the former and minimise the latter as the technology develops. Rejecting a technology because it has some problems is nonsensical. Activists who promoted the slogan "ban the bomb" were asking for the impossible, since we cannot uninvent something or turn back history.

We should also recognise that problems might never be

solved, but have to be managed. Nuclear war is still possible, perhaps started by a tyrannical regime like the ones in North Korea and Iran. The devastation might not be as much as an all-out war between superpowers, but many millions could die. Politicians in democracies need to adopt policies that encourage certain countries not to develop nuclear weapons, or deter them from ever considering using them.

> **Discussion: What other inventions have good points and bad points, and how have we maximised the former and minimised the latter? Consider the motor car, or the internet.**

ENVIRONMENTAL IMPROVEMENTS IN MY LIFETIME

When I was four years old, my father took me to a hill overlooking Glasgow at the end of July after the factories had been closed for the fortnight trades holiday. My father knew that it was the only time of the year when we could obtain this panoramic view of the whole of Glasgow and much of central Scotland to the mountains beyond, because for the rest of the year it was shrouded in smoke from industrial pollution. Gradually, during the 1960s, '70s and '80s, the environment was cleaned up as coal power was substituted with less dirty fuels like oil and gas, especially after plentiful North Sea reserves were discovered. Rivers in urban areas which had been running sewers for industrial waste started to support stocks of fish for the first time in over a century.

When I was five years old, my father had accumulated enough wealth to purchase a house in the countryside just beyond that hill, and I grew up observing nature and farming practice. By the age of nine, I travelled back into the city on a steam-powered train to attend school, and the view of urban desolation remaining from the Industrial Revolution inspired me to become an architect. James Watt first developed steam engine technology in Glasgow and a quarter of the world's steam engines were made in the city. Europe's first ocean-going steam ship went into service on the River Clyde and its estuary. But times had moved on, and Glasgow had to reinvent itself.

And so started my interest in urban development and the natural world, with farming practice capable of supporting both. A favourite holiday destination was the Spey Valley in the Scottish Highlands, an area which has become the UK's largest national park. A single osprey pair had built a nest at Loch Garten which the RSPB protected and made available for the public to view from hides. Having been extinct in the UK for a century, there are now more than a hundred breeding pairs. Similar conservation measures to protect sea eagles and red kites have almost gone unnoticed in a wealthy country which can afford to protect large areas of its natural world, while allowing access to humans to enjoy it. Deer used to be illusive creatures that my father would take me on long car journeys to search for, but they are now commonplace in lowland areas, and have become so numerous that their numbers have to be controlled. Venison was once a rich man's food but is now readily available in supermarkets.

I applauded Greenpeace in its early days when it campaigned for an end to atmospheric atomic bomb tests (the Test Ban Treaty was signed in 1963), the clubbing of baby seals to death by fishermen to protect their stocks in the Arctic (conservation measures were introduced in 1973) and an end to whale hunting (only partly successful with some countries stopping the practice in 1969 and a more extensive ban on commercial whaling in 1986). The brave and adventurist activists got in the way of ships at sea to shield whales from the harpoons of fishermen. I was less impressed by the long haired, bearded, sandal-wearing hippies who stayed at home and protested to democratic governments, when the sources of the problems were in other countries. They could easily be ignored as moaning protesters looking for others to solve their problems, not action men (and women) who made a real difference, proactivists rather than just activists. I could not understand CND's campaign to have nuclear weapons banned because it is impossible to uninvent a technology, and their objective could only have been achieved by setting up a dictatorial, undemocratic world authority to impose such a strategy. Placing that power in a centralised authority would be very dangerous for many reasons.

WORRIES ABOUT CLIMATE CHANGE – TOO COLD OR TOO WARM?

I first came across climate change when I read Nigel Calder's 1974 book *The Weather Machine*, which explained the various mechanisms that led to the weather we experience. Calder was editor of *New Scientist* magazine from 1962 to 1966 and was much respected. In the book he noted that the world's climate had been getting cooler from about 1940, and various eminent scientists were speculating that we could be heading back to a new Little Ice Age. This might happen very quickly because of various possible tipping points which would accelerate the phenomenon. He could not know that, a few years later, the world would start to warm up again, because we have to wait a few decades to observe a new climatic trend.

The concept of anthropogenic global warming crept up on me over many years, as it did for most people. We were aware that fossil fuel reserves might not last forever, and that we should seek alternatives at some point. While the first British wind farm opened in 1991 in Cornwall, it was several years later that wind turbines became more than a curious idiosyncrasy. I was annoyed when a wind farm was built on a beauty spot on a hillside viewed from Stirling Castle. Was there nothing sacrosanct? When my architectural practice designed a new headquarters for Scottish Natural Heritage in 2006, it achieved its status as the most environmentally sensitive office building in the UK up to that time due mainly to its energy efficiency. Wind and solar power were not viable, either practically or financially. The next year we considered biofuels to power a new hospital development, but rejected them because no one could guarantee the supply of materials over the 30-year life of the energy system. Also, the huge number of heavy goods vehicles constantly delivering the volume of materials required to power a large, high-tech hospital was unviable, and would have led to a huge amount of CO_2 emissions.

The UK Climate Change Act of 2008 seemed to pass through without any public debate, and it was only after I retired in 2014, when I attended a conference where a green activist referred to "climate chaos" that I appreciated how much of a passion this subject is to many people. I had experienced many extreme weather events in my seven decades on the planet, and there did not appear to be anything unusual about what was happening now. This spurred me on to research the subject in as much detail as I could. This book is the result. I could not claim to be a 'climate scientist' but, building on the knowledge gained from two science-based university degrees, I have practised the scientific method throughout my career to much success, with energy efficiency a key principle in the many schools, colleges, hospitals, offices, shopping centres and residential buildings that my architectural practice designed. In my research I also found that the so-called 'climate scientists' I spoke to relied on very questionable evidence for their conclusions.

POST SECOND WORLD WAR FEARS ABOUT HUMANITY'S POWERS

I could not see 'climate chaos' or a 'climate emergency', and my research on why people should believe this took me to the origins of environmentalism which had passed me by in the 1960s and '70s. In 1962, the year of the Cuban Missile Crisis, Rachel Carson's book *Silent Spring* said that harmful chemicals associated with agriculture were poisoning wildlife and would lead to widespread cancer in humans. Birds starved of bugs which were being poisoned would become extinct and the countryside would become silent. Like much doomsday cult thinking, some of it arose from legitimate concerns about poorly tested new pesticides and herbicides being applied indiscriminately, but some of it lacked a sound scientific basis, relying on worst-case scenario speculation and fears about point-of-no-return outcomes.

> "Only within the moment of time represented by the present century has one species – man – acquired

significant power to alter the nature of the world. During the past quarter century this power has not only increased to one of disturbing magnitude but it has changed in character. The most alarming of all man's assaults upon the environment is the contamination of air, earth, rivers and sea with dangerous and even lethal materials. This pollution is for the most part irrecoverable; the chain of evil it initiates not only in the world that must support life but in living tissues is for the most part irreversible. In this now universal contamination of the environment, chemicals are the sinister and little-recognised partners of radiation in changing the very nature of the world – the very nature of its life. Strontium 90, released through nuclear explosions into the air, comes to earth in rain or drifts down as fallout, lodges in soil, enters into the grass or corn or wheat grown there, and in time takes up its abode in the bones of a human being, there to remain until his death. Similarly, chemicals sprayed on croplands or forests or gardens lie long in the soil, entering into living organisms, passing from one to another in a chain of poisoning and death. Or they pass mysteriously by underground streams until they emerge and, through the alchemy of air and sunlight, combine into new forms that kill vegetation, sicken cattle, and work unknown harm on those who drink from once pure wells. As Albert Schweitzer has said, 'Man can hardly even recognise the devils of his own creation'."

A new religion had been born. Man's "chain of evil" had "created" "devils", "alchemy" and "contamination". Armageddon was inevitable because his sinful and "sinister" actions had "mysteriously" caused "universal", "irrecoverable" and "irreversible" "lethal" consequences. We were all doomed. Except we were not. We adapted by carefully testing new fertilisers, pesticides and herbicides and producing regulations to make sure that their harm was minimised and their benefit maximised.

The fear factor led to the ban of the chemical DDT in

1972, despite the US Environmental Protection Agency's nine-month hearing concluding that "DDT is not a carcinogenic [cancer causing] hazard to man… The use of DDT under the regulations involved here does not have a deleterious effect on freshwater fish, estuarine organisms, wild birds or other wildlife." DDT had been key to killing mosquitos and eradicating malaria in southern USA and the south of Europe, but conditions of western aid to African countries now precluded it. The rich countries which had benefitted from DDT thought they now knew better and would not allow its use in poorer countries. This was an instance of *Green Power, Black Death*, a book by Paul Dreissen, where millions of lives could have been saved, but were lost.

Gradually, as many African nations relied less on aid, they turned to DDT as a tool in their fight against malaria. When I worked for the South African health authorities in 2013 I found that they used it extensively. DDT should have been regulated, not banned. Believing it was doing good, but failing to observe scientific principles, the overreaction of green politicians caused unintended consequences and great harm to vulnerable people. 'Do-gooders' in rich countries seeking to control the lives of people in poorer countries is called 'neo-imperialism'.

> **Discussion: It is right that rich countries should give aid to poorer ones. But should they tell the poor countries how to spend the money or just leave them to use it in the way they think best suits their circumstances? Should activists be people who take positive action, as Greenpeace used to do, not just produce placards, cause roadblocks and march outside government buildings?**

DEATH FROM THE SKY

Then there was the scare about the 'hole' in the ozone layer. By the mid-1970s there was a concern that human-produced chlorofluorocarbons (CFCs) and halons (HCFCs), present in refrigerants, solvents and aerosol propellants, and transported into the stratosphere by winds, might be accelerating the breakdown of ozone (O_3) into oxygen (O_2) in this natural oxygen-cycle process. The ozone layer prevents the most harmful ultra-violet band (UVB) wavelengths from passing through the atmosphere from the Sun.

In 1985 it was reported that the ozone column over Antarctica in September and October had dropped by up to 70%, and the hysteria about the 'hole in the ozone layer' began. The concept of a hole in the atmosphere allowing death rays to cause skin cancer for all humanity was scary enough for Americans to stop buying aerosol sprays before legislation was passed in 1978 to ban CFCs. There never was a 'hole' in the ozone layer, just a 'depression' for a few months each year over an uninhabitable region, but 'depression in the ozone layer' would not have had the desired political scare effect. Concerns led to the Montreal Protocol in 1987 to ban worldwide the production of CFCs, halons and other ozone-depleting chemicals. It was easy to find alternatives like hydrocarbon refrigerants, roll-on deodorants and pressure sprays. The politicians claimed to have saved us all from ecological disaster, but we were just being sensible and choosing alternative technologies with less potentially harmful effects. Poor countries were oblivious to the change because they did not have electricity to power refrigerators anyway, nor could they afford deodorants.

EXHAUSTING THE WORLD'S LIMITED RESOURCES

The doom merchants have regularly predicted that we would use up the world's resources, and the civilised life that depends on these would end. At university, around 1970, I was told that world supplies of copper would run out within a few decades, and that economic development of India and China would be stopped because communications technologies relied on copper. My lecturer could not have predicted that fibre optic cables and Wi-Fi would be invented, and by 2020 there would be more copper reserves than ever before. In theory, the world's natural resources must be limited, but human knowledge and resourcefulness seem unlimited. Experience has shown that we develop more efficient techniques to make the most of what we have, or we find alternatives.

There was a similar story about oil. Sure enough, US production peaked in 1970 at 10 million barrels per day and declined to a low point of 4 million barrels per day in 2008. However, thanks to hydraulic fracturing techniques, by 2017 production was again at more than 10 million barrels per day, with America passing Saudi Arabia and Russia as the top oil producer and Russia as the biggest gas producer. America uses more hydrocarbons than in 1970, but it should be a net oil exporter within a decade. The Green River Formation in Colorado alone contains up to 3 trillion barrels of untapped shale oil, half of which may be recoverable, five and a half times the reserves of Saudi Arabia.

THE FEAR OF TOO MANY PEOPLE

Then there was the panic about 'overpopulation'. Paul Ehrlich's 1968 book *The Population Bomb* predicted mass starvation as the world's population rapidly increased. Well-publicised famines in the 1960s and '70s fueled the concern, but these were almost always caused by wars and civil unrest. Ehrlich was spectacularly wrong because he did not predict that food production would rise faster than population due to nitrogen fertilisers and new varieties of corn. We used science to solve our problems. These trends will continue with GM (genetically modified) technologies such that more crops can be grown in less space with less fertiliser, less pesticides, less herbicides and less soil erosion. What should be a thoroughly positive outlook is constantly eroded by the merchants of doom.

Chapter 7 shows how population growth stops when people are wealthy enough to plan families, and yet organisations like Population Matters, with prominent members David Attenborough and fellow British naturalist Chris Packham, still contend that there are other humane ways to reduce population growth, despite overwhelming evidence to the contrary. I say 'humane' ways, because dictators Stalin, Hitler, Mao and Pol Pott showed how to kill tens of millions of people by inhumane means.

It seems that many nature documentary makers 'go native' and place wild creatures' needs over those of humans. In 1991, in an interview with the *UNESCO Courier*, underwater explorer and TV personality Jacques Cousteau said: "What should we do to eliminate [human] suffering and disease? It's a wonderful idea but perhaps not altogether a beneficial one in the long run. If we try to implement it we may jeopardise the future of our species… It's terrible to have to say this. World population must be stabilised and to do that we must eliminate 350,000 people per day. This is so horrible to contemplate that we shouldn't even say it. But the general situation in which we are involved is lamentable." When the Duke of Edinburgh was a patron of WWF he said, "If I were reincarnated, I would wish to be returned to Earth as a killer virus to lower human population levels."

Incidentally, Ehrlich initially hedged his bets on climate change since the world had been in a cooling phase from about 1940, saying, "With a few degrees of cooling a new ice age might be upon us," but, "with a few degrees of heating the polar ice caps would melt, perhaps raising ocean levels 250 feet." Then, in 1986 he said: "As University of California physicist John Holdren [later an advisor to Barack Obama] has said, it is possible that carbon dioxide climate-induced famines could kill as many as a billion people before the year 2020". Another hero of the green movement got it hopelessly wrong.

KEEP CALM AND CARRY ON – HAVE A PROPORTIONATE RESPONSE TO PROBLEMS

There have been many other scare stories in my lifetime, like acid rain and large-scale soil erosion, but they either did not happen as badly as some predicted, or we found ways to alleviate or stop them. We will undoubtedly face problems again, whether due to human actions or merely natural occurrences, and they will usually be unpredictable, like COVID-19. I cannot guarantee that we will get on top of them, but we always have in the past, so I remain confident that we will do so again. We naturally fear what we do not know, and tend to exaggerate problems to encourage others to take them seriously, but we need to be careful not to take steps that will cause more harm than good. The cure must not be worse than the disease. For example, as we shall learn later, biofuels are far from carbon neutral and their production is very harmful to the environment.

> **Discussion: What predictions about what will happen next week, or next year, would you be happy to make? What factors make it impossible to predict the future?**

MY CURRENT FEARS

I suppose my current fear is that we are over-estimating the problem of climate change and, in our desire to address it quickly, we are adopting inappropriate and harmful measures. It is only rich countries which are taking the issue seriously, and they can probably afford to make the mistakes so that others can learn from them. The main downside occurs when we demonise fossil fuels, despite them being necessary for many decades to come, primarily because wind and solar do not work most of the time. So we have to import oil and gas from countries with nasty regimes like Saudi Arabia and Russia. We have to kow-tow to them instead of condemning them outright. The economies of Germany and Eastern European countries rely on Russian gas supplies such that they are still under the influence of their former Soviet oppressors.

Democratic countries like Britain and Germany have adequate supplies of gas which could be obtained using modern extraction techniques, and have plentiful supplies of coal which has no affordable substitute for powering steel furnaces. But the power of the green movement has made this impossible. The inevitable result has been high and uncompetitive energy costs, such that we transfer industry to China, which also has an oppressive regime. Democracies are sustaining autocracies by weakening their own economies and, by importing fossil fuels and manufactured goods instead of obtaining them locally, are increasing emissions from unnecessary transportation.

We like to think that science drives our way of thinking,

and we justify our approach to anything like climate change by claiming the authority of scientists (or at least the ones who reinforce our beliefs), as if they are infallible and not subject to the prejudices of colleagues or the politics of the organisation they work for. But we are emotional creatures who are as likely to fall subject to popular myths as to irrefutable truth. The Enlightenment philosopher David Hume (1711–1776) turned traditional religious and secular logic on its head by saying, "Reason is, and ought to be, the slave of the passions… Reason's role is purely instrumental; it teaches us how to get what we want. We are, in the end, creatures of habit, and of the physical and social environment within which our emotions and passions must operate. We learn to avoid the passions which destroy, and pursue the ones that succeed. Society has to devise strategies to channel our passions in constructive directions."

At the moment we are a slave to the emotions of the greens, who have the unimpeachably honorable aspiration to 'save the planet'. I have met many passionate climate campaigners who live in Western democratic countries which afford the most benign, safe, privileged, healthy social conditions that have ever existed. There are imperfections in our societies that we should strive to correct, but they are rarely life-and-death issues. By focusing on the worst-case scenarios of climate change, and the part that human emissions may play, they have found their existential cause, and it is no use reasoning with them, or pointing out the flaws in their theories and policies, because passion trumps logic.

As someone who believes in the inherent value of liberal democracy, my nightmare is the prospect of an authoritarian elite believing that it has the answer to humankind's problems, and imposing it on us. The Club of Rome was formed by two very rich and influential people: Aurelio Peccei, an anti-fascist resistance fighter in the Second World War who rose to be president of Olivetti; and Alexander King, a Glasgow-born research scientist who was credited with harnessing DDT against lice and mosquitos during the war to save millions of lives, but which he later seems to have regretted. They met in the mid-1960s and reckoned that the planet was overcrowded, resources needed to be conserved, economic growth was the problem and not the solution, and urgent action must be taken via a pan-global authority, preferably the UN. Later strategy envisaged this to be funded by carbon taxes, and economic growth would be stopped by decarbonising economies. This policy failed to obtain approval at the Copenhagen Summit in 2009.

The 1972 book *The Limits to Growth*, which sold ten million copies, set out the Club of Rome's agenda. It started the trend of producing computer models to 'prove' desired outcomes. The most famous Club of Rome statement is, "The common enemy of humanity is man. In searching for a new enemy to unite us, we came up with the idea that pollution, the threat of global warming, water shortages, famine and the like would fit the bill. All these dangers are caused by human intervention, and it is only through changed attitudes and behaviour that they can be overcome. The real enemy then is humanity itself."

It gets even scarier. "Democracy is not a panacea. It cannot organise everything and it is unaware of its own limits. These facts must be faced squarely. Sacrilegious though this may sound, democracy is no longer well suited for the tasks ahead. The complexity and the technical nature of many of today's problems do not always allow elected representatives to make competent decisions at the right time." In other words, elected politicians are too stupid to understand issues relating to the dangers involved in economic growth. The powers that utopian environmentalists seek to obtain would not be dissimilar to the fascist ones that Pecci fought against. As Christopher Booker once said, "Evil men don't get up in the morning saying 'I'm going to do evil'. They say, 'I'm going to make the world a better place'."

My hope, however, is that common sense will win out in the end, because the truth has a way of revealing itself at some point, and people have to admit that they got it wrong. Key to this is the desire of 6 billion people on the planet to obtain the living standards of the richest nations. They will make common sense judgements rather than ephemeral, aspirational ones. As Hume's contemporary Adam Smith (1723–1790) wrote: "The natural effort of every individual to better his own condition is so powerful a principle that it is not only capable of carrying on the society to wealth and prosperity, but of surmounting a hundred impertinent obstructions with which the folly of human laws too often encumbers its operation." If Hume and Smith were alive today, they might cite Ed Miliband's 2008 Climate Change Act as the perfect example of this folly.

CHAPTER 3
WHAT DOES THE UNITED NATIONS SAY ABOUT CLIMATE CHANGE?
METEOROLOGY AND FUTUROLOGY LESSONS

Sorry, this is starting to get very technical but, trust me, you need to know in some detail what the United Nations says about the subject. Many people only read the summary at the beginning of each assessment report it produces, but in my comments I give more technical details. As ever, the devil is in the detail. This chapter is long, but it is essential to understand the topic properly.

The Intergovernmental Panel on Climate Change (IPCC) was set up by the UN Environment Program (UNEP) and the UN World Meteorological Organization (WMO) in 1988 to investigate the link between human activities, primarily the emission of greenhouse gases, and changes in the climate, fundamentally global warming. It produced five major Assessment Reports (*IPCC AR 1–5*) in 1990, 1995, 2001, 2007 and 2013, as well as other reports on related issues. In 2018 it produced a special report on the impacts of global warming of 1.5°C (compared to 2°C), sometimes referred to as *IPCC SR1.5*. As its name suggests, it is a political organisation which reviews the work of thousands of scientific studies to draw conclusions and formulate reports. While I will be critical of its work in certain respects, and it is not the final word on climate change, it is a good starting point. It is certainly the most authoritative source of information on the subject.

I will concentrate on two reports mentioned above, those

from 2013 and 2018, on which the basis of my quest was founded. *IPCC AR6*, published in August 2021, came after I had written most of this book, so is reviewed in an appendix. The results in an IPCC report are generally not definitive conclusions, but assessments of the likelihood of certain conditions taking place or predicted to take place in the future. For example, *IPCC AR5* says that it is 95% certain (*IPCC AR4* was 90% certain) that most of the global warming since 1950 has been caused by human greenhouse gas emissions. It does not say if the human contribution is 51%, 71% or 91%, presumably because our understanding of the sensitivity of greenhouse gases on the climate system is still not well understood.

However, most activists must believe that it is towards the last of these because they do not differentiate between 'anthropogenic' (human produced) global warming (AGW) and overall global warming. They assume that climate change is effectively human-induced, and natural climatic changes play a small or insignificant part in what we are observing. To activists, the words 'climate change' are synonymous with 'anthropogenic climate change' as if only human actions cause change, and therefore our remedial actions will return the climate to some ideal natural average world temperature. IPCC reports do not talk about a 'climate emergency' or 'climate chaos', so we must be careful to note what they say, without the exaggeration that activists use. Their conclusions are as follows, summarised here under various headings. I have made comments on each, eleven in total, and then make general conclusions.

GLOBAL WARMING

From around 1850 until 2018, the average global temperature rose by 1°C ±0.2°C; in other words, somewhere between 0.8°C and 1.2°C. The temperature graph up to the time of *IPCC AR5* is in Figure 7. In the northern hemisphere, 1983–2012 was "likely" the warmest 30-year period of the last 1,400 years. *IPCC AR5* noted that the rate of warming in the 15 years from 1998 to 2012 slowed to 0.04°C/decade (UK Met Office graph HadCRUT4), effectively zero, as against the 0.11°C/decade average from 1951

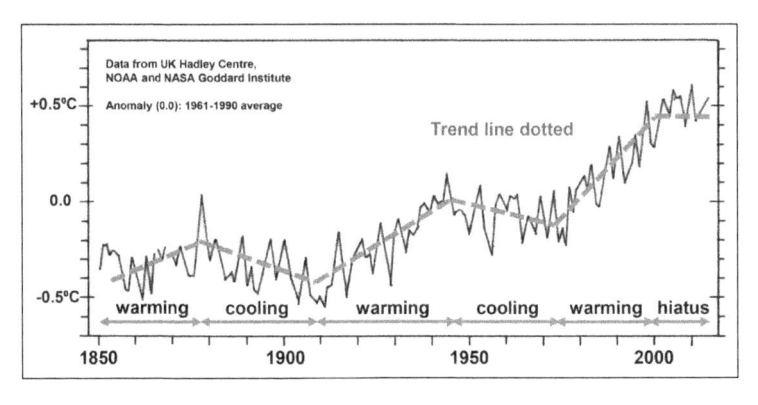

Figure 7 Global average temperatures 1850–2013.

to 2012. This pause in the warming, or "hiatus" as it was called, was further discussed on pages 769 to 772 of *IPCC AR5* with three possible reasons for it:

1. There was warming but it was lost within the overall climate system (such as in the deep oceans);
2. The radiative forcing of the Sun varied (the Sun radiated less heat – unlikely); or
3. There was a mistake in the models (they "show a larger response to greenhouse gases than in the real world").

Comment 1: The world has certainly warmed up since 1850, but it is surprising how broad the estimate is, somewhere between 0.8°C and 1.2°C. The IPCC does not say why this range is so large, but I can explain. For a start, historic records of temperatures, apart from a few locations, are very poor. Ian Harris at the Climate Research Unit at the University of East Anglia played a part in compiling data from around the world. When he was working on input from the Australian Bureau of Meteorology he noted: "What a bloody mess. Now looking at the dates… something bad has happened, hasn't it? COBAR AIRPORT AWS cannot start in 1962, it didn't open until 1993! … getting seriously fed up with the state of the Australian data. So many new stations have been introduced, so many false references… so many changes that aren't documented… I am very sorry to report that the rest of the databases seem to be in nearly as poor a state as Australia was… Aarrggghhh! There truly is no end in sight!" When you read in a newspaper that a record temperature has been set in Australia, remember how poor the historic records are.

Even in recent decades, there has been a big margin of error in arriving at a maximum and minimum temperature for each day (an average between the two is then calculated) for each 100 km x 100 km grid box that the world is divided into for the convenience of computer models. Each reading is as much estimated as measured, because:

- The temperature will vary significantly in parts of each 10,000 km^2 box, although this is a small problem because we are interested in the 'anomaly' or change from a defined average temperature over time, not the actual temperature.
- There are large parts of the world where we do not have weather stations, so the temperatures are noted for the grids closest to the gaps, and estimates made.
- Many weather stations are now in urban areas and gain heat radiated from hard, man-made surfaces. This is called the Urban Heat Island effect. A good example is the Sydney observatory which used to be in parkland but is now on a traffic island in a ten-lane motorway system with multi-storey buildings all around it. Each day someone must guess what the temperature would have been without the UHI. The US Senate General Accounting Office reported in 2011 that 42% of its stations failed to meet the correct siting standards. Anthony Watts' studies reckon this figure is even higher.

Comment 2: *IPCC AR5* says that some parts of the world were as warm as present during the Medieval Warm Period (called the Medieval Climate Anomaly in *IPCC AR5* and defined as between the years 950CE and 1250CE) but that not all were, thereby justifying the statement that the last 30 years were the warmest since about the year 600CE; in other words, the end of the Roman Warm Period. The Medieval Warm Period was followed by the Little Ice Age, which ended around 1820, the low base from which the current 1°C rise started. It would not be good to go back to such a cold period, but climate activists say that a rise of one or two degrees from this baseline would be bad. The thousand-year cycle of natural warming and cooling periods is discussed further in Chapter 12.

A painting of the River Thames in London in 1677 by Abraham Hondius, at the height of the Little Ice Age. © *Heritage Images*

Comment 3: It is often said that climate sceptics made up the pause in warming between 1998 and 2014, but *IPCC AR5* recognised it. While temperatures have risen since 1950 (after the global cooling period 1945–75 in Figure 7), the rise has not been consistent, nor has it matched the shape of graph of increasing human emissions. This surely questions the certainty of human emissions being the dominant factor in warming.

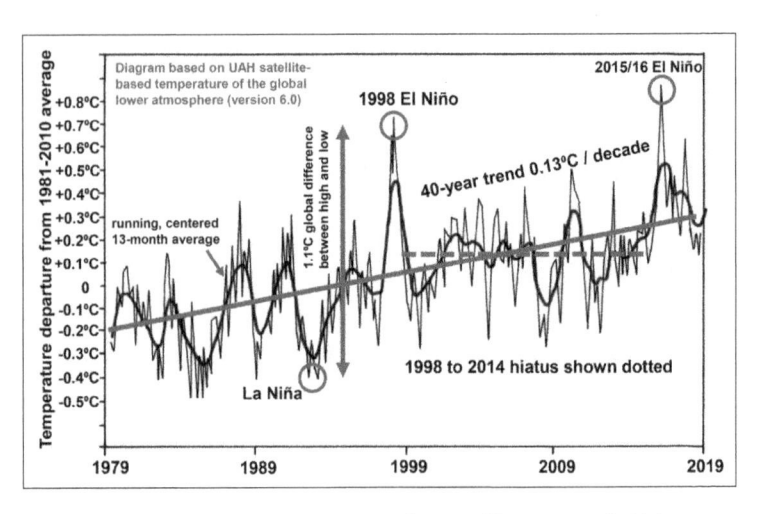

Figure 8 Satellite-based temperatures. Source: University of Alabama in Huntsville.

Given the problems with recorded land surface temperatures, many scientists prefer to trust satellite measurements of temperature in the lower atmosphere, where the weather happens and which rises above, and evens out, the hot spots caused by urban heat islands. This is shown in Figure 8, from NASA satellite data compiled at the University of Alabama in Huntsville by researchers John Christy and Roy Spencer. The pause in the warming is clear between the two El Niño high points in 1998 and 2015/16 (dotted line). The warming rate for the 40 years since this data has been collected is 0.13°C per decade, similar to the IPCC's estimate of 0.11°C/decade for 1951 to 2012. This is not an 'alarming' rate of warming. Notice also how a natural event like an El Niño high and a La Niña low can vary the average world temperature by around 1.1°C over a period of about five years. The rate of rise is 0.17°C/decade in the northern hemisphere and 0.1°C/decade in the southern hemisphere because the north is mainly land and the south mainly ocean.

The IPCC had to say that "1983–2012 was 'likely' the warmest **30-year period**" because the definition of climate change is an observed trend over at least this length of time, but it admits that there was a pause in the warming from 1998. In fact the two periods of warming in the 20th century, roughly the 1920s and '30s and the 1980s and '90s, were both less than 30 years' duration so, technically, there was no period of climate change during this century. This book could end here with the contention that modern climate change is a myth, but a technical knock-out will not suffice.

POLAR AND GLACIAL ICE

Over the last two decades, the Greenland and Antarctic ice sheets have been losing mass, glaciers have continued to shrink almost worldwide, and Arctic sea ice and northern hemisphere spring snow cover have continued to decrease in extent. "The Greenland and Antarctic ice sheets have decreased in mass. Data from NASA's Gravity Recovery and Climate Experiment show Greenland lost an average of 286 billion tons of ice per year between 1993 and 2016, while Antarctica lost about 127 billion tons of ice per year during the same period. The rate of Antarctic ice mass loss had tripled in the last decade.

"With 1.5°C of global warming, one sea ice-free Arctic summer is projected per century. This likelihood is increased to at least one per decade with 2°C global warming (*high confidence*). It is *very likely* that the annual mean Antarctic sea ice extent increased at a rate in the range of 1.2% to 1.8% per decade between 1979 and 2012." There is less spring snow. "Satellite observations reveal the amount of spring snow cover in the northern hemisphere has decreased over the past five decades and the snow is melting earlier."

Comment 1: One hundred and twenty-seven billion tonnes of lost Antarctic ice sounds a lot but it is only 0.00048% of the ice sheet, whose volume is difficult to accurately calculate. In computer model inputs, a small change in the assumption of average density per cubic metre in each layer of ice can make a big difference when applied to the entire Antarctic ice continent of 26,500,000,000,000m^3. Jay Zwally, a NASA glaciologist, reckons that the ice in West Antarctica is reducing, while the larger main mass of ice elsewhere, all on land, is increasing. West Antarctica extends north over the Antarctic Circle into warmer waters. Andrew Shepherd of the University of Leeds obtained slightly different results for the last 25 years, with the conclusion that East Antarctica could also be losing ice. Both work on slightly different assumptions for the density of the ice. Researchers writing in *Nature Climate Change* in July 2020 identified the Interdecadal Pacific Oscillation, with ocean current changes, as the mechanism which increases or decreases Antarctic ice on a 15- to 30-year cycle. In other words, there is a natural explanation for these changes.

Because the Antarctic is such a hostile place (coastal temperatures vary from minus 3°C in summer to minus 26°C in winter, while inland it varies from minus 32°C in summer to minus 68°C in winter), historic data is limited, but we can see from Figure 9 the routes that early explorers took, and the ice extent now is similar.

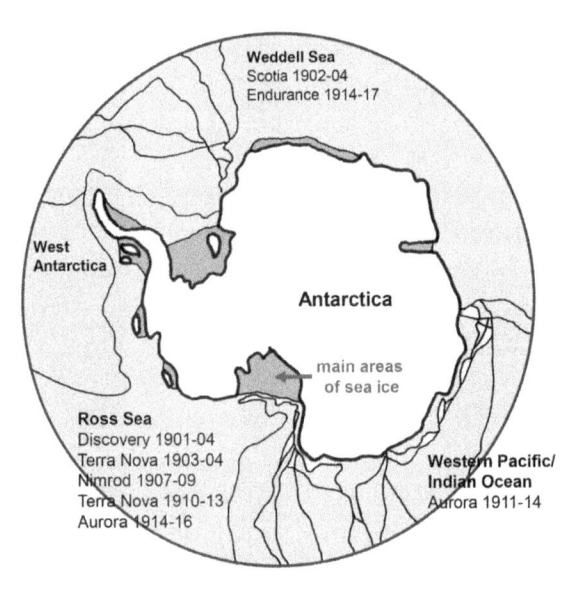

Figure 9 Expedition routes taken by explorers between 1897 and 1917. Sources: *The Cryosphere* and British Antarctic Survey.

Comment 5: Arctic ice floats on the ocean and since about 1980 it has seen a downwards trend. The general warming over the last century may well be having a long-term effect, but judging the ice extent "over the last two decades" could be misleading. There is other evidence which suggests that there is a 60-year cyclical element in Arctic sea ice extent, possibly related to a cycle of ocean currents. There was a North Atlantic ocean warming period from about 1975 to 2006, which appears to be reversing (see Figure 10). This will be explored in more detail in Chapter 12. *IPCC SR1.5* confirms that the long-term trend is slow, with only a 1-in-100-year likelihood of no summer ice when the globe warms up from the present 1°C above pre-industrial temperatures to 1.5°C, perhaps about the middle of the 21st century.

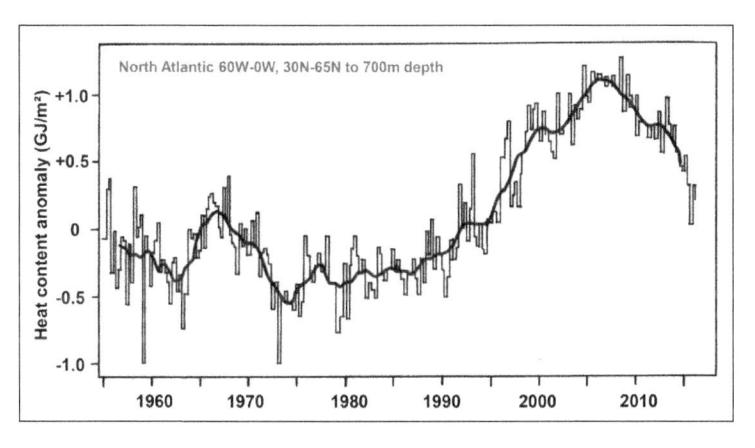

Figure 10 North Atlantic heat content 1955–2017.

Taking short-term trends, and assuming that they will continue, explains why so many predictions about the disappearance of Arctic ice have been wrong. On 20 June 2008, *National Geographic* quoted David Barber of the University of Manitoba, from aboard the Canadian research icebreaker CCGS Amundsen, "We're actually predicting this year that the North Pole may be free of ice for the first time." On 24 June 2008, the Michigan *Argus Press* reported NASA Goddard Institute's James Hansen, "'We see a tipping point right before our eyes. The Arctic is the first tipping point and it's occurring exactly the way we said it would." Echoing work by other scientists, Hansen said that "in five to ten years, the Arctic will be free of ice in the summer". The ice did not disappear the way he said it would by 2018. On 14 December 2009 Al Gore said that, "Arctic sea ice may disappear by summer 2014". It is still there, despite the eminence of the people predicting its absence, and it actually peaked in 2014 (Figure 11).

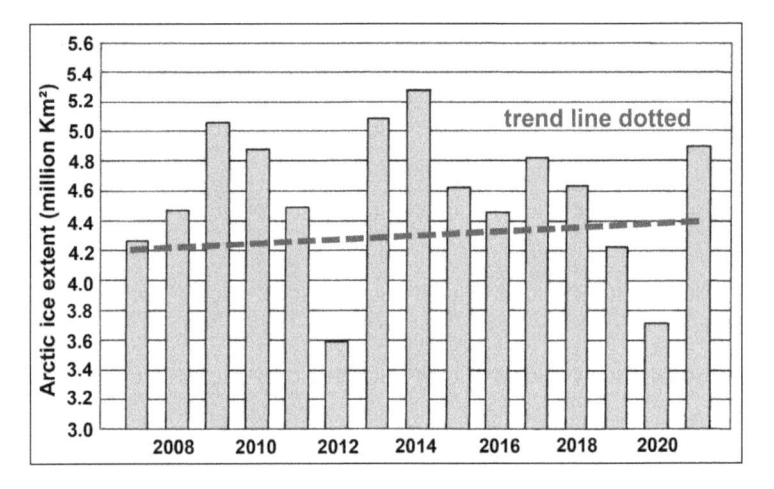

Figure 11 Arctic September sea ice extent 2007–21.

Figure 11 charts the September Arctic sea ice extent, the time when it is at its minimum, from 2007 to 2021 to demonstrate a slightly rising trend for this period, but with significant annual peaks and dips. (Notice that the base of the vertical scale is 3 million square kilometres, so even when it is at its lowest there is still a lot of ice.) The datasets are the Sea Ice Index (SII) from NOAA and the

Multisensor Analyzed Sea Ice Extent (MASIE) from NIC, but they are almost identical. Is this merely a pause in the downward trend or a change in the trend altogether? This is the problem with climate change. We need to wait until 20 to 30 years have passed to confirm if a perceived trend has actually happened.

Comment 6: During the period from about 1980, when Arctic sea ice was reducing, *IPCC AR5* confirmed that Antarctic sea ice was increasing. The world cannot be taken as a whole, with different climatic conditions happening in different locations due to a variety of natural factors.

Comment 7: Less snow and ice alters the 'albedo effect' of the planet's surfaces, with less 'whiteness' meaning less reflection and more absorption of the Sun's rays, leading to even more warming. *IPCC AR5* mentioned that spring snow cover has been decreasing. It did not mention that the database it uses, compiled at Rutgers University, noted a corresponding increase in autumn snow cover, with the overall annual snow cover showing no trend (Figure 12). Perhaps cyclones hold more water and produce more snow when they hit colder air as winter approaches, but this is only a guess. We don't know for sure.

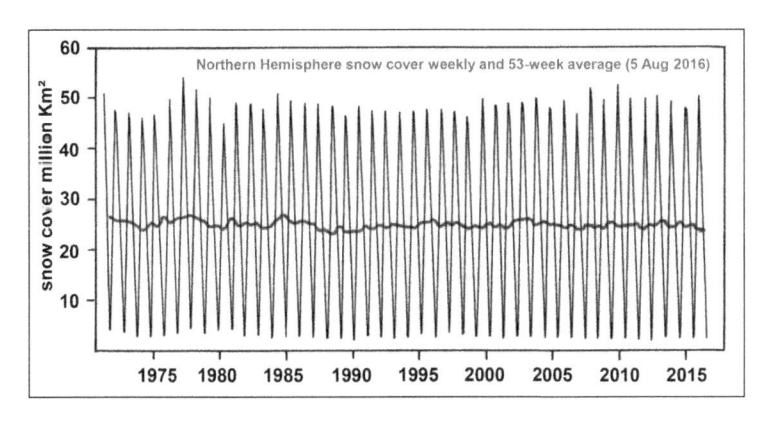

Figure 12 Northern hemisphere annual snow cover 1972–2016.

SEA LEVELS

"Global sea level rose about 8 inches [200 mm] in the last century. The rate in the last two decades, however, is nearly double that of the last century and is accelerating slightly every year. It is *likely* that similarly high rates occurred between 1920 and 1950."

Comment 8: Rising sea levels are obviously a potential threat to coastal communities, and a large proportion of the world's population lives on coastal plains. Many low-lying islands might disappear completely. But the rate of sea level rise has been slow and, even if there has been a short-term acceleration to 320 mm/century "in the last two decades", *IPCC AR5* points out that this also happened between 1920 and 1950. A slow rise allows sea defences to be strengthened in most areas that are under threat. A lot of Holland is below sea level.

Again, there have been dramatic predictions based on short-term trends. In 1996, James Hansen told a reporter that within 20 years: "The West Side Highway [beside the Hudson River in New York] will be under water and there will be tape on the windows across the street because of high winds; and the same birds won't be there. The trees in the median strip will change. There will be more police cars [because] you know what happens to crime when the heat goes up." But by 2016 The Battery Tide Gauge showed the same slow rise in sea level (284 mm/century) since 1850 (Figure 13).

Other locations also show a steady rise with only short-term variations off the trend, for example, in the UK where *Woodworth et al.* 2009, taking data from stations in Aberdeen, North Shields, Sheerness, Newlyn and Liverpool, recorded a trend of 140 mm/century since 1900. Perhaps this steady rise has been an inevitable result of emerging from the Little Ice Age, with the huge volume of water in the oceans moderating any shorter-term variations in the atmospheric warming process, like the warming phases in Figure 7 of

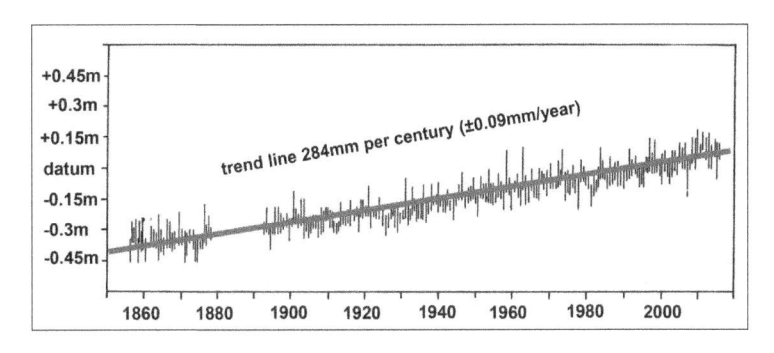

Figure 13 Sea level trend in New York 1850–2017. Source: NOAA.

1850–80, 1920–45 and 1975–98, and the cooling/pause phases of 1890–1920, 1945–75 and 1998–2015.

Seas are rarely flat but form swells and uplifts when there is varying atmospheric pressure, or the gravitational influence of continents and ice sheets. The sea levels either side of the Panama and Suez canals are different, which surprised engineers when they were built. An El Niño event changes the level of the Pacific Ocean by at least a metre. The sea expands when it gets warmer, but the land can also go up and down over time. The north of the UK is slowly rising as it springs back from the Ice Age 20,000 years ago when it was under 3 kilometres of heavy ice.

It is often said that Bangladesh is in danger of losing land to rising sea levels, but the huge amount of sediment swept down the Rivers Ganges and Brahmaputra from the Himalayas is increasing the land area of this country faster than any likely sea level rise could reduce it. All of these natural variations make it hard to detect human factors.

Some believe that Pacific island nations will disappear as rising seas overwhelm them. President Macron at the One Planet Summit in December 2017: "Behind me are the heads of state and governments. In 50, 60, 100 years, there are five, ten, fifteen who won't be there." On 17 May 2019 UN Secretary-General António Guterres said, "We must stop Tuvali from sinking and the world from sinking with Tuvali."

And yet the scientific evidence says something different. As Charles Darwin knew, large storms break off coral from reefs and pile it up on islands. A 2015 study in *Geology* magazine by *Kench et al.* of the Funafuti Atoll showed a 7.3% increase in net island area over the past century, and a February 2019 University of Auckland study of Tuvali, using aerial photographs and satellite images between 1971 and 2014, showed that the island had grown by 2.9%. Paul Kench's most recent results, from January 2021, reveal that only 10% of the Marshall Islands, Kiribati and the Maldives have reduced in area in the last six decades. Most had increased.

CHANGES IN THE CLIMATE

It is "likely" that the frequency of heat waves has increased in large parts of Europe, Asia and Australia, and there are "likely" more land regions where the number of heavy precipitation events has increased than where it has decreased, mainly in North America and Europe. There is "low confidence" of increases in intensity and/or duration of drought on a global scale, although the changing pattern of climate means that there will always be some areas where rainfall patterns change periodically. There is also "low confidence" in the long-term increase in intensity of tropical cyclones.

Comment 9: It makes sense that there will be more heatwaves and flash floods when the atmosphere is warmer, but we have the ability to install air conditioning in homes and offices, and engineer rivers so that they only flood on flood plains. The UN confirms that droughts are unlikely to increase globally (as Figure 3 shows) and cyclones (storms) are unlikely to increase in intensity or frequency (as Figures 4 and 5 suggest). When you see news reports about individual storms and activists say that they are getting more frequent and wilder, remember that we need several decades of statistics to draw conclusions, and these tell us that a warmer planet does not seem to produce cyclones that increase in frequency or intensity.

HUMAN INFLUENCE ON THE CLIMATE

The atmospheric concentrations of carbon dioxide, methane and nitrous oxide have increased to levels unprecedented in at least the last 800,000 years. Carbon dioxide concentrations have increased by 40% since pre-industrial times, primarily from fossil fuel emissions and secondarily from land use change. The ocean has absorbed about 30% of the emitted anthropogenic carbon dioxide, causing ocean acidification.

Human influence on the climate system is clear. This is evident from the increasing greenhouse gas concentrations in the atmosphere, positive radiative forcing (greenhouse gas radiation – see Chapter 8), observed warming and understanding of the climate system.

Human influence has been detected in warming of the atmosphere and the ocean, in changes in the global water cycle, in reductions in snow and ice, in global mean sea level rise and in changes in some climatic extremes. This evidence for human influence has grown since *IPCC AR4*. Human influence has been the dominant cause of the observed warming since the mid-20th century.

Comment 10: Whether changing forests into farmland or emitting greenhouse gases, it is certain that 7.6 billion human beings have affected the climate. The debate is about how much, and whether there is anything significant we can do about it, short of reducing the human population back down below the one billion that lived before the Industrial Age. Because there is not a close correlation between the way the temperature has risen in steps (Figure 7) and the straight line rise in atmospheric CO_2 (Figure 23 in Chapter 8), nor between the upwards curve of emissions and the straight line rise in atmospheric CO_2, I cannot see how we can be sure that humans should take the lion's share of the blame. I also cannot conclude that it is a bad thing for the world to be a little warmer (1°C more than the Little Ice Age). What may be worrying is the likelihood of a continued warming trend towards the end of this century and beyond.

One of the reasons given for believing that humans have caused the warming is that the UN's advisors have so far not come up with a better explanation. In Chapter 9 I will say why some scientists believe that the answer lies in a better understanding of the water cycle, specifically the part that clouds and oceans play in moderating the radiated heat from the Sun, as it goes through cycles of intensity. Their arguments seem rational and credible. I will explore ocean acidification in Chapter 5.

PREDICTING THE FUTURE WITH CLIMATE MODELS

Four scenarios are run on the computers for the period from 2013 to 2100, with the possible temperature rise ranging from a low of 0.3°C to a high of 4.8°C (the mean of each scenario is between 1°C and 3.7°C), and the possible global mean sea level rise ranging from 0.26m to 0.82m (the mean of each scenario is between 0.4m and 0.63m).

As it says on page 824 of *IPCC AR5*, "Climate models are extremely sophisticated computer programs that encapsulate our understanding of the climate system and simulate, with as much fidelity as currently feasible, the complex interactions between the atmosphere, ocean, land surface, snow and ice, the global ecosystem and a variety of chemical and biological processes... Climate models of today are, in principle, better than their predecessors. However, every bit of added complexity, while intended to improve some aspect of simulated climate, also introduces new sources of possible error... Furthermore, despite the progress that has been made, scientific uncertainty regarding the details of many processes remains."

Comment 11: This large range of possible temperature and sea level outcomes seems inevitable, given the margin of error that is present in such a complicated system, due to the limited accuracy of the data that we feed in (like temperatures), and our imperfect understanding of how the various elements of the climate system react with each

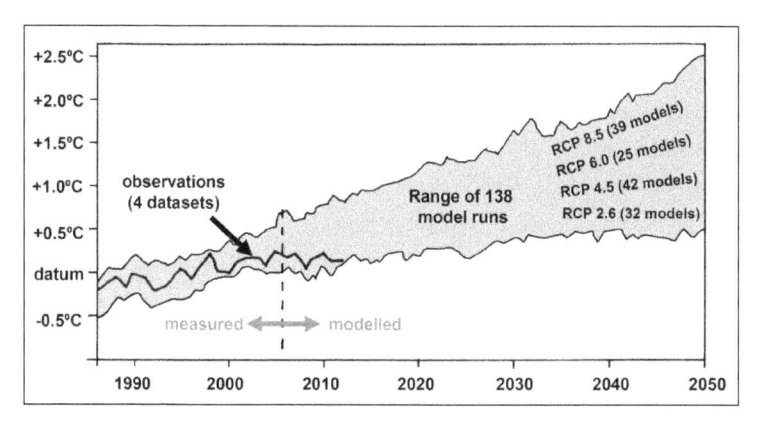

Figure 14 Chart based on page 981 of *IPCC AR5* with 138 predictive models compared to observations.

other. The IPCC talks of the "butterfly effect", where a small change or error in the input to a computer model can be amplified exponentially to result in a huge inaccuracy or error in the output. The more complex the model, the more chance of this, and yet, if the model is too simple, it misses out on vital factors. This is a dilemma.

It makes it easy to argue, on the one hand, that there is a slow-burning fuse that we have time to put out, or that it is fast-burning and we must immediately blow very hard on it. The current estimated rate of anthropogenic global warming (no natural warming/cooling is considered) is between 0.1°C and 0.3°C per decade. As we have seen, the rate of warming for the last 40 years has been about 0.13°C/decade, and *IPCC AR5* itself contains a graph which shows that the actual warming has been at the bottom of the range of predictions (Figure 14) since about 2000, possibly because the pause in warming from 1998 to 2015 was not predicted. Some scientists believe that the warming will gradually catch up with the modelling, but that is merely speculation. The warming rate is at the lower end of the possible range, which is of concern for the future, but not alarming.

CONCLUSIONS

If we ignore what the activists say and only pay heed to the above UN IPCC reports, there is cause for concern about global warming in the medium to longer term, that is from about 2040 to the end of the century. It seems that we are heading for a temperature rise of between 1.5°C and 2°C above the pre-industrial temperature by later in this century, with the possibility that some, so far undetected, feedback in the system or "tipping point" might accelerate this to perhaps 3°C or more.

But there are a small number of scientists who reckon that we have entered a cooling phase in solar activity (the Sun switches polarity about every 11 years, leading to 22-year solar cycles) and that there may be some natural reduction in the rate of warming for a few decades, as there was on two occasions in the 20th century and from 1998 to 2015. The UN has been trying to persuade governments that they need to take climate change seriously, and it is understandable that it will not wish to take its foot off the pedal, but it seems we may have time to invent new, alternative technologies to help us reduce our emissions, if we need to. There is certainly no need to panic.

Discussion: Does my logic make sense? Does it help to explain just how complicated the climate systems are, or does it just lead to confusion? Why has this information not been presented before in a book addressed to ordinary people, especially young people? Why do people make predictions several decades into the future when they will almost always be wrong? Why are they always predictions of doom, rather than of good times to come?

CHAPTER 4
TIPPING POINTS AND CLIFF EDGES
FUTUROLOGY LESSON

Chapter 3 concludes that human beings are playing a part in global warming but, while this may need to be addressed going forward, we do not have a good enough understanding of all the factors involved in climate change to declare that we need to take drastic measures at this time, especially if they are expensive and ineffective, as I will show in Chapter 10 that many are. My analysis would change if the warming suddenly accelerated for any particular reason, so I need to present and discuss what factors might cause this.

This is best illustrated in an article by Tim Lenton of the University of Exeter in *Nature* on 27 November 2019, where he outlined seven possible tipping points (he first identified these in 2008 in the *Proceedings of the National Academy of Sciences*). Coincidentally, *The Guardian* newspaper of the same date announced the article (this must have been pre-planned for maximum effect), with Phil Williamson of the University of East Anglia saying, "The prognosis by Tim Lenton and colleagues is, unfortunately, fully plausible that we might have already lost control of the Earth's climate." This supposes that humankind ever had control of the world's climate in the first place, a concept which seems strange.

It is clear who Lenton is targeting in *The Guardian* article. "We might already have crossed the threshold for a cascade of interrelated tipping points. The simple version is the schoolkids are right: we are seeing potentially irreversible changes in the climate system under way, or very close. As a scientist, I just want to tell it how it is. It is not trying to be alarmist, but trying to treat the whole climate change problem as a risk management problem. It is what I consider the common sense way."

In fact, Lenton is not telling it 'like it is' but speculating about what might happen in a very unscientific way (futurology is not a branch of science). As he says himself, his arguments stem from limited understanding of the systems involved, and are imaginative guesses. "Some scientists counter that the possibility of global tipping remains highly speculative. It is our position that, given its huge impact and irreversible nature, any serious risk assessment must consider the evidence, however limited our understanding might be. To err on the side of danger is not a responsible option. If damaging tipping cascades can occur and a global tipping point cannot be ruled out, then this is an existential threat to civilisation. No amount of economic cost-benefit analysis is going to help us. We need to change our approach to the climate problem." Scientists should not have a "position"; they should merely present facts, and we cannot afford to respond to every "highly speculative" future scenario that imaginative people can contemplate.

The seven possible tipping points are below, in italics, with my comments following.

1. *The West Antarctica ice sheet may be in reversible retreat.*
    But it might not. We have not observed this long enough to detect trends. This ice sheet extends north beyond the Antarctic Circle into warmer waters, so it is subject to periodic melting from variable ocean currents.

    Lenton's timescale for melting ice is not days, or even years, or even decades, or even a few generations. "A **model study** shows that when this sector [the Amundsen Sea embayment of West Antarctica] collapses, it **could** destabilise the rest of the West Antarctica ice sheet like **toppling dominoes** – leading to about 3 metres of sea level rise on **a timescale of centuries to millennia**. Palaeo-evidence shows that such widespread collapse of the

West Antarctica ice sheet has occurred repeatedly in the past." I have highlighted key words. "Toppling dominoes" suggests speed, but the effects are on "a timescale of centuries to millennia", and of course the scenario is a construct of a "model study" world. Humanity could cope with a 3-metre sea level rise in 1,000 years.

2. *The Greenland ice sheet is melting at an accelerating rate.*
   It has been in recent decades but before this it increased in depth. Abandoned Second World War aircraft were recently found under 60 metres of ice. We cannot draw long-term conclusions from short-term statistics.

3. *The Arctic sea ice is shrinking fast.*
   Ice shrank from the 1980s to about 2007 but prior to this it increased, with all predictions of ice-free summers being revised as they are proved wrong. Figure 11 does not indicate a fast rate of decline.

4. *The Gulf Stream has slowed by 15% since 1950*
   But the article itself says that this is within the range of natural variability. Lenton also speculated that melting fresh water from the Arctic and Greenland will further slow the Gulf Stream, a scenario that was considered possible in Nigel Calder's 1974 book *The Weather Machine*, but this time caused by global cooling, and dramatised in the 2004 science fiction film *The Day After Tomorrow*, where the conditions accelerated over a period of days to plunge New York into an ice age.

5. *Seventeen per cent of the Amazon rainforest has been lost since 1970, and a tipping point is predicted to take place at between 20% and 40% loss which will lead to the forest "drying out".*
   But most of the forest was lost in the 1980s and 1990s and experience shows that increases in farmland in developing countries level off as modern, efficient

agricultural practice is adopted. It is possible that the 20% figure will not be reached, but the fact that there is such a large range in the prediction (20% to 40%) reflects its highly speculative nature.

6. *Global warming leads to more fires and pest outbreaks in temperate forests, especially in North America, turning some regions from a carbon sink into a carbon source.*
Yet satellite records show that areas affected by wildfires have been reducing, we know that better technology allows us to deal with pests better and probably a bigger threat to North American forests is the destruction of vast areas for biofuel exports to Europe.

7. *In the tropics, corals will be wiped out by 2°C of warming.*
But we know that corals adapt to slow temperature changes, thrive in warmer waters, and are only temporarily bleached by quick, El Niño-induced local temperature rises of about 5°C. There is more about this in the next chapter.

8. *As temperatures rise, permafrost, mainly in Canada and Russia, melts and releases methane which has been stored in the ground, further accelerating global warming.*
But this proves that temperatures have been warmer than present in the past when this vegetation grew, rotted and gave off methane, and that previous natural warming periods did not have sufficient power to cause tipping point events that stopped subsequent natural cooling from placing the land back under permafrost.

It is easy to dream up catastrophic events from a small amount of evidence. For example, a team of scientists from Edinburgh University found that there are 91 volcanoes under the West Antarctica ice sheet which could cause enormous disruption if some erupted, although they are under 2 km of ice and their activity is unknown. This team did the proper scientific thing by merely presenting the facts and not speculating about doomsday scenarios, despite the attention and possible future research funding this might attract. But eruptions would be natural events which Lenton could not blame humanity for. Melting ice caused by ocean warming is a better bet because it has a tenuous link, via a warmer atmosphere, to human emissions, but it is happening so slowly that it needs a theoretical tipping point to make it significant.

I first came across Tim Lenton when I read his interesting 2011 book *Revolutions that Made the Earth*. He wrote it with Andrew Watson of the University of East Anglia. This traced the development of the planet over 4.6 billion years, including the formation of conditions that supported life. This happened through a number of revolutionary stages, or steps, brought about through tipping points, but on a geological timescale. Stages such as the moon forming, ocean formation, the formation of the oldest rocks, snowball Earth stages, carbon isotopes (3.7 billion years before present – BP), conditions allowing life to start (3.5 billion years BP), photosynthesis (2.7 billion years BP), the Great Oxidation (2.3 billion years BP), the evolution of animals (metazoa, from 1.8 billion years BP), multi-cellular life (900–600 million years BP), plants (470 million years BP), mass exterminations (the first two being End-Permian 251 million years BP, End-Cretaceous 65 million years BP) and ice ages.

In reading the book, I had been wading through eons of changes to planet Earth, learning about them in great technical detail and, keen to get his understanding of anthropogenic climate change, I expected similar detail on this. I was disappointed that he covered the topic in about one page, with little more than the following. "To our minds the combination of the well-developed theory for the greenhouse effect, and the observed temperature increase, makes the case for this causal connection unequivocal. There is an unprecedented level of scientific consensus about it, and there should be no need to debate the matter further, but for some powerful vested interests seeking to promote the appearance of uncertainty and lack of consensus. The climate is actually changing faster than many existing model projections, which have used 1990 as a start."

First of all, as I showed in Figure 14, the IPCC itself recognises that model projections overestimate climate change. Why does he believe the opposite of this? He falls back on scientific consensus, without giving any evidence of this, and wants to shut down debate. He is suggesting that anyone who would want a debate must have sinister "vested interests". My vested interest is getting to the truth, and if people like Lenton cannot explain in some detail how AGW works, or what the "scientific consensus" is, in a book similar to this, I will naturally question his scientific objectivity.

I can understand why some academics get involved in politics, because they rely on the political process to obtain public research funding. But as soon as they enter the political arena, they become biased and say unscientific things like "consensus". To quote James Delingpole: "Let's be clear: the work of science has nothing to do whatever with consensus. Consensus is the business of politics. Science, on the contrary, requires only one investigator who happens to be right, which means that he or she has results that are verifiable with reference to the real world. In science consensus is irrelevant. What is relevant is reproducible results. The greatest scientists in history are great precisely because they broke the consensus."

I can also understand how academics are motivated by political initiatives. The IPCC tends to set the agenda and research studies follow it. There were only a handful of research papers with the words 'climate change' in them produced each year in 1987, before the IPCC was formed, but this rose to around 3,000 each year by 2003. After Al Gore's film, *An Inconvenient Truth*, the graph steepened again to a peak of about 12,000 per year in 2010. The figure has now reached 40,000 in this field where 'the science is settled'.

But much of Gore's contentions and predictions were blatantly wrong. Some educationalists in Britain wished to circulate his film to schools but this was stopped by a successful court action which highlighted nine assertions that were not scientifically justified.

- Sea level will rise by 20 feet
- Pacific atolls are being inundated
- The ocean conveyor is shutting down
- CO_2 levels and a rise in temperatures are an exact fit
- Mount Kilimanjaro's snow recession is attributable to global warming
- Global warming is drying up Lake Chad
- Global warming caused Hurricane Katrina
- Polar bears are drowning because of global warming
- Coral reefs are being bleached by global warming

My second experience of Lenton came when I enrolled for an online course on climate change in 2018 with the University of Exeter. He and colleagues gave various lectures on a number of aspects. There was a chat line for students on which I recorded my reservations on the current level of science, and apparent contradictions in the established theories. Why are we worried about a 1°C rise in temperatures from the Little Ice Age? Why has the warming over the last century taken place in a series of steps which do not correlate with the rise in emissions? Why are predictive climate models mostly wrong? Surely we have a very poor understanding of the sensitivity of the climate to greenhouse gases? I received no comments from any of the lecturers, and concluded that I was being subjected to what was effectively preaching and propaganda, not scientific dialogue.

And yet, while Lenton is "unequivocal" about the link between CO_2 and warming, in 2011 he was also candid about our poor knowledge on "what the 'climate sensitivity' to increased CO_2 is. The heat-trapping properties of water vapour at high concentrations are not very accurately known, and neither are the properties of clouds in such a steamy atmosphere. So while we know there is a cliff edge out there, that the Earth is moving very slowly towards, we aren't sure just how far away it is… In the unlikely event that doubling CO_2 increases temperatures by much more than the 3°C that is expected, we should be very alarmed… As far as we can see no one has yet made a convincing scientific case that we are close to a global tipping point for 'runaway' climate change."

His September 2019 article gives the impression that a tipping point may be imminent, rather than at some point in the far future, but it doesn't say what new compelling evidence has changed his mind, and why tipping points previously thought to be possible with a 4°C or 5°C temperature rise are now thought possible at 2°C or 3°C. Is this just cranking up the alarm in concert with the increasingly dire warnings from the IPCC because practically none of the world's countries are adopting policies towards a zero-carbon economy?

He expects a doubling of CO_2 to cause around 3°C of warming because, "The warming effect of the increasing burden of greenhouse gases is amplified, by roughly a factor of two, by an equivalent contribution from increased water vapour in the atmosphere." Figure 15 shows the key interactions and feedback loops on the planet as Lenton sees them, but assumes that water vapour only gives positive feedbacks (increasing warming) and not negative ones (decreasing warming) by increased clouds reflecting sunlight.

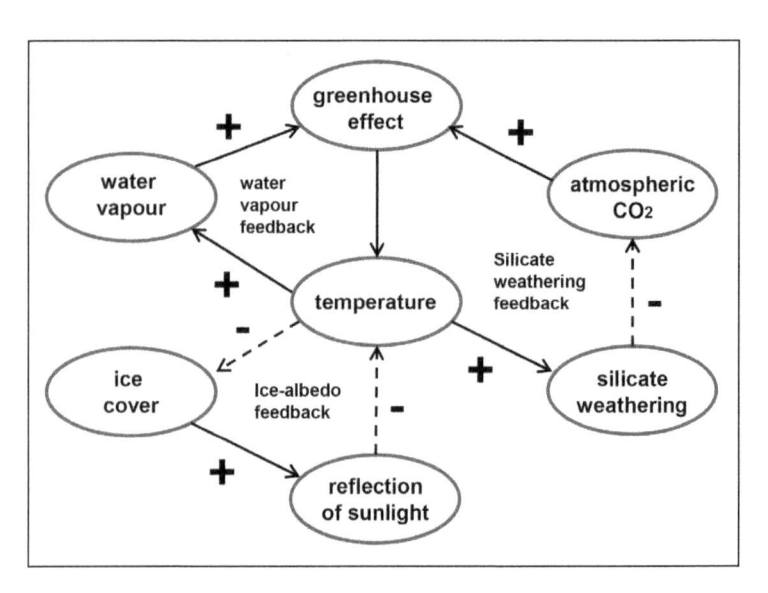

Figure 15 Key climate change feedbacks. Source: Lenton.

Lenton also understands the limitations of modelling. "We've been at pains to point out whenever we are in danger of believing in the model too literally, that **it is just a toy**. It is useful because it captures, indeed **exaggerates, a particular aspect** that we are interested in – the influence that a small number of rare but important events could have on the evolution of the Earth system. **It's not the real system** however, but a **caricature** of it and much is left out that is important. In particular, we have barely discussed the reorganisation of the whole Earth system – the 'revolutions' of this book – because those transitions are more complex events, that can't be readily fitted into the constraints of the model." In other words, models miss out important aspects, emphasise the bits that the modellers are most interested in, and cannot be used to construct complex events like tipping points.

I cannot question Lenton's technical knowledge, intelligence and humanity, but I can criticise him for choosing to cross the Rubicon between the theoretical scientific researcher and the common sense, real-worldly policy player, ignoring inconvenient gaps and inconsistencies in the data as he goes. He is qualified and well suited for the former role but not the latter, like hundreds of other academics who promote the climate change agenda. I am giving him a hard time here, but I will show in Chapter 12 how easy it is for people in his position to fall into the trap of allowing science to follow politics, and for experts to work in a bubble of consensus, or 'groupthink' as Irving Janis, the former Yale professor of psychology, defined: "Those caught up in a bubble first succumb to a collective mindset which is in some way at odds with reality. They then elevate this into an illusory orthodoxy which cannot be challenged. Finally, because their groupthink is based on such shaky ground, they intolerantly lash out at anyone who dares question it."

In trying to determine the facts on the subject, I am very aware that no one academic is an expert in the range of disciplines necessary to understand what the climate is doing or might do, so each one is little more equipped than any educated layperson who has studied various aspects of climate change over many years, like me. Perhaps because a layperson is more detached from the internal politics within each discipline, he/she can retain their scepticism better. It was physicist Richard Feymann who said, "Science is the belief in the ignorance of experts."

In this book, by presenting what appear to me to be the facts, I was hoping to alleviate eco-anxiety, but I am generating another fear – that of bad science taking us down poorly thought out policy routes. The IPCC encouraged a lot of research focused on proving that humans affect the climate, and people like Tim Linton persuade us that we do not need to understand the science ourselves because there is scientific consensus (I have searched for this but cannot find it). The politicians, believing they are doing good, then thrust policies on us without any public debate, justifying these in statements like the following very first words from a March 2020 UK government consultation paper: "Climate change is the most pressing environmental challenge of our time. There is overwhelming scientific evidence that we need to take action." I contend that none of the people reading that document had any real understanding of the relevant science, and I believe that the consequent lack of debate is undermining the democratic process.

> **Discussion: How easy is it to fall in with the crowd and try to please everyone else by conforming to the group view? What examples of this have you experienced, but have felt uncomfortable about? How would research scientists obtain funding if they did not follow the wishes of government policy? When would you whistleblow on an employer, or would you just stay quiet?**

Tim Linton argues that if the effects of climate change might conceivably be catastrophic, we should act to mitigate them; in other words, drastically reduce human emissions. However, we have no way of knowing what the consequences of tipping points are (because we have never experienced them before), how likely they are to happen (if at all) and whether reducing emissions will have any effect (our understanding of the sensitivity of the climate to emissions is poor). We might just be spending a lot of money, wasting a lot of time and making a lot of people suffer from a poorer standard of living, for nothing. A professional risk manager would not adopt such a strategy.

> **Discussion:** Consider six possible risks you might face getting to school/college/work on time. Have a wide range from ones that might be more likely (the morning alarm fails to go off) to some that are very unlikely (being caught up in a terrorist attack). Consider what chance each risk has of happening, what the consequences might be and what steps (mitigation measures) you might take to stop them happening. Some will be easy to prevent, some might have such small consequences that they are not worth bothering about, while others might be so unlikely, or unpreventable, or the mitigation might be more troublesome/costly than the problem, that we would just deal with them if they happened, when they happened, using our personal resourcefulness. Where do you think climate change fits within such an assessment?

CHAPTER 5
ANOTHER PANIC: OCEAN ACIDIFICATION
ECOLOGY AND CHEMISTRY LESSONS, HIGHLIGHTING THE IMPORTANCE OF GOOD QUALITY CONTROL IN RESEARCH

THE ALARMIST AGENDA
In his 2007 book *Six Degrees: Our Future on a Hotter Planet*, Mark Lynas started with a quote from Dante's *Inferno*, where Dante enters the First Circle of Hell: "From the weeping ground there sprang a wind, flaming with vermillion light, which overmastered all my senses, and I dropped like a man pulled down by sleep." Global warming of perhaps 6°C is predicted to have hellish consequences for us all and for the natural world.

If humans are increasing the amount of CO_2 in the atmosphere, a lot of it will be absorbed into the oceans. This should make the water less alkaline, or more acidic, and it might react with the calcium in the shells of sea creatures. Lynas quoted Ken Caldeira of the Carnegie Institution's global ecology department: "In less than 100 years, the pH of the oceans would drop by as much as half a unit from its natural 8.2 to about 7.7." A neutral pH level is 7.0, which is neither alkaline nor acidic, so a drop of 0.5 from 8.2 (1.2 above neutral) is almost half of its alkalinity lost; in other words, he speculated it would get almost twice more acidic.

Lynas went on: "Acidification will directly affect ocean creatures. Crabs and sea urchins need their shells to survive, whilst fish gills are extremely sensitive to ocean chemistry, just as our lungs are to air. Mussels and oysters, vitally important both as economic resources and as part of coastal ecosystems worldwide, will lose their ability to build strong shells by the end of the century, and will dissolve altogether if atmospheric CO_2 levels ever reach as high as 1,800 ppm."

Ocean acidification has been described as an impending 'ocean calamity' and the 'evil twin of global warming', but we do not have clear historic measurements on which to understand natural variability and consider long-term trends. Monthly data from the Hawaii Ocean time-Series (HOt) programme only started in 1988. Most research is only a few decades old. Measurements of ocean acidification in *IPCC AR5* are only quoted from three observatories over a period of two decades, where the pH dropped from about 8.12 to 8.08. It is somewhat mischievous for it to claim that this small sample over a short timescale is good enough to draw conclusions. The sole source of the claim in *IPCC AR4* that climate change was linked to coral reef degradation was a Greenpeace report called *Pacific in Peril*. One report by a highly biased activist organisation is not enough.

And human emissions could not push atmospheric CO_2 levels anywhere near to 1,800 ppm because populations, energy use and emissions level off when a country's economy reaches that of the West (see Chapter 7), and should do so for the other 6 billion people who will give up high emissions (and high pollutant) coal for gas power as their economies develop. Even if economies and populations did continue to expand, for the last 40 years atmospheric CO_2 has been rising at a fairly constant 1.5 ppm/year (despite rising emissions), so at this rate it would take 800 years to get to 1,800 ppm.

Shellfish need CO_2 from the ocean to make their calcium carbonate shells which, when they die, fall to the seabed, forming sediments. The chalk we see today in the white cliffs of Dover was laid down during the Cretaceous Period between 145 million and 66 million years ago when there was an abundance of shellfish for one fundamental reason. Atmospheric CO_2 was over 2,000 ppm (Figure 16 in the next chapter) and creatures used this dissolved gas in the sea to build their shells. Demonising this gas, which is also a building block for all photosynthesising plants, and calling it a pollutant, is as illogical as damning the oxygen that animals breathe.

THE STATE OF CURRENT RESEARCH

In 2004 there were no scientific papers published with the words 'ocean acidification' in them. By 2010 there were about 250 papers being produced worldwide and this rose to a peak of around 700 each year by 2013. Howard Browman, the editor of *ICES Journal of Marine Science*, and a research scientist at the Norwegian Institute of Marine Research, reckoned that the bias in favour of doom-laden articles on this subject was partly the result of pressure on scientists to produce eye-catching work. "You won't get a job unless you publish an article that is viewed as of significant importance to society. People often forget that scientists are people and have the same pressures on them and the same kind of human foibles. Some are driven by different things. They want to be prominent... The oceans will never become acid because there is such a huge buffering capacity in them. We simply could never release enough CO_2 into the atmosphere to cause the pH to go below 7. If they had called it something else, such as 'lower alkalinity', it wouldn't have been as catchy."

John Abbot and Jennifer Marohasy provided a useful summary of the most prominent research projects in the book *Climate Change: The Facts 2017*, published by the Australian Institute of Public Affairs. According to them: "As is true across all of science, studies that report no effect of ocean acidification are typically much more difficult to get published and, if published at all, seem to appear in lower-ranking journals. These are subsequently ignored by the mainstream media."

HOW 'ACID' ARE THE OCEANS?

At present the ocean pH varies around the globe between 7.5 and 8.4, and constantly changes depending on ocean currents, storms, undersea volcanos (there are many more of them than terrestrial ones), erosion of rocks, outfall from rivers and many other factors. *Kline et al.* 2015 showed that at Heron Island on the Great Barrier Reef the pH generally falls below 8.0 soon after midnight in summer, while climbing to about 8.4 in the afternoon in winter. *Revelle & Fairbridge* 1957, before the current concerns about acidification were born, reported pH levels of 9.4 in isolated coral reef pools during the warmth of the day, falling to 7.5 at night. The explanation for this is that, at night, sea organisms continue to respire (breathe out) CO_2, while there is no uptake of it through the mechanism of photosynthesis by plant life (like seaweed) as there is during the day when the sun shines. Organisms have evolved to cope with these variations.

There are no historic records of ocean pH, but certain sources claim that the world average has dropped from 8.2 to 8.1, will drop to 7.8 by 2100, and that this must be due to the extra CO_2 produced by humans. Such calculations are mostly the result of modelling or laboratory experiments, not observations in the more complex real world. Approximate historic pH values have been estimated using boron isotopes in coral in the Flinders Reef on the Great Barrier Reef (*Pelejaro et al.* 2005) and in the South China Sea (*Liu et al.* 2009), which indicate that a local average range of 7.9 to 8.3 is normal on a century- or thousand-year scale.

THE DIFFICULTIES IN RESEARCHING CARBONATE CHEMISTRIES IN THE OCEAN

One of the seemingly most comprehensive studies, which gained much prominence and caused a lot of concern, was of 328 colonies of Porites corals from 69 reefs in the Great Barrier Reef by *de'Ath et al.* 2009. *Ridd et al.* 2013 reviewed this and found a systematic data bias in the last growth band of each sample core which erroneously concluded a recent drop in calcification (coral forming) rates. Abbot and Marohasy: "Most ocean acidification research is focused on a single species with investigations into their short-term physiological response. Understanding a whole-of-ecosystem response, however, often requires some understanding of the relative impacts on different species, including competitive and trophic (feeding) interactions."

They highlighted the Free-Ocean CO_2 enrichment (FOCe) experimental approach, carried out in-situ with partially open enclosures and a through-flow of seawater. Using this, *Georgiou et al.* 2015 showed that, "the corals maintained their calcifying fluid pH at near-constant elevated levels, independent of the highly variable temperatures and FOCe-controlled carbonate chemistries to which they were exposed. This is an important finding because it implies that corals may have a high degree of tolerance to ocean acidification."

This is complicated, with a lot of technical jargon, so let me summarise. Studies on a single species of coral can be misleading, especially if they are carried out in a laboratory, because there are various species and other organisms which interact with each other in a dynamic way. The FOCe experimental approach takes place in the ocean where all these factors are present, and it is deemed more reliable. It showed that corals are tolerant to changes in the alkalinity/acidity of the water.

The corals look harmed because they bleach (go white) when more sensitive organisms (microscopic algae) that live on them are rejected by the coral polyps as the water temperature changes (up or down). But when the water temperature stabilises, the algae return, or different algae inhabit the reef, a process which physical oceanographer Peter Ridd has observed since 1985. Bleaching, like wildfires on land where plant life quickly grows back, is a natural process which corals have evolved to respond to and recover from. Corals grow quicker in warmer waters, so a slow increase in temperatures of perhaps 1°C over several decades is a good thing. Long-term atmospheric global warming, whatever the human contribution is to this, cannot account for coral bleaching.

WHAT THE US NATIONAL OCEANIC AND ATMOSPHERIC ADMINISTRATION SAYS

A lot of fuss has been made about the bleaching of large sections of the Great Barrier Reef, and that ocean acidification must be to blame. NOAA issues a factsheet on the subject. It gives four reasons for bleaching, none of which mention acidification:

1. Change in ocean temperature
2. Run-off and pollution
3. Overexposure to sunlight
4. Extreme low tides

It does suggest that climate change contributes to ocean temperatures, but we know that corals adapt to slow changes in temperature, and they live in areas with warmer water in the Arabian Gulf than in Australia. The bleaching events seem to coincide with El Niño events where the water temperature changes rapidly (up to 5°C) over the period of a few years; but they have recovered within a relatively short time before, the microscopic algae returning to give them colour again. If the sea level is rising with global warming, this is good for corals because Reasons 3 and 4 would come into play if it were dropping.

As a result of a Freedom of Information request, a memo was released dated 25 September 2015 from Dr Busch of NOAA's Ocean Acidification Program and NW Fisheries Science Center in Seattle, and it said: "Currently there are no areas of the world that are severely degraded because of OA [ocean acidification] or even areas that we know are definitely affected by OA right now." *Nature* magazine in May 2017 stated that corals' ability to make skeletons is "largely independent of changes in seawater carbonate chemistry, and hence ocean acidification". That seems pretty conclusive.

BLEACHING EVENTS IN THE GREAT BARRIER REEF

A number of alarming studies coincided with the warming western Pacific Ocean due to the large 2015/16 El Niño event, where predictions of large-scale death of the Great Barrier Reef were made. Bleaching, which was at its peak from January to April 2016, rarely leads to coral death, and Peter Ridd's caution was proved correct, with only 8% of the reef dying, from which recovery can take a few years. Ridd knew that corals grew in warmer waters, so it was not the temperature itself that was causing problems, but the relatively rapid change in water temperature caused by an El Niño. Cyclone Debbie in April

2017 also contributed significant physical damage and sediment cover to some sections of the reef. Indonesian studies also blamed the change in sea level during the El Niño due to the western Pacific dropping by about a metre in level, exposing more coral at mean low spring tide level where they normally thrive. The differentials in related atmospheric pressures between Tahiti and Darwin have been recorded since 1876.

A January 2016 article in *Nature* claimed that there were no longer any living colonies of Acropora, a type of coral, at Stone Island off Bowen Harbour. In 2019 Jennifer Marohasy, Walter Starck and others found a 100% healthy coral cover in this location. Their study, called *Beige Reef*, is on YouTube. Bleaching events do not often lead to widespread coral death, but the emotive, apparently lifeless pictures grab headlines. They are just part of the natural cycle of reef life.

To me the most convincing, common sense argument for questioning whether more dissolved CO_2 is causing problems for the Great Barrier Reef is what I call 'The Coke Test'. Human beings love fizzy (carbonated) drinks where CO_2 is forced into liquids to enhance the taste, the bubbles helping to stimulate our taste buds. Warm, flat Coke is horrible. CO_2 dissolves best in liquids (or oceans) between 5°C and 8°C, so we keep our Coke in the fridge (at about 5°C) and put ice in it to increase the time that the bubbles stay when subjected to a room or outdoor temperature that could be in excess of 20°C. If ocean acidification were to happen, it would be noticed first in cool oceans, not warm ones like those off north-eastern Australia.

One of the areas with the lowest (natural) level of pH is Patagonia, where sea life flourishes in the colder, lower alkaline water, influenced by the Humboldt Current. In *The Wizard and the Prophet*, Charles C Mann recounted the experience of American ornithologist William Vogt (the environmentalist prophet of doom of the title) when studying cormorants on the Chinca islands off the coast of Peru in 1939 at the beginning of an El Niño event. The water temperature rose by about 10°C over a period of months, altering the way nutrients were distributed by the current and lowering the CO_2 level of the water, which drastically reduced the amount of phytoplankton, which deprived the cormorants of the anchovies and sardines they lived on. Five million birds disappeared within a month, leaving behind their chicks to starve to death. The birds had still not returned in 1941. It is understandable that people were concerned by the effects of the 2016 El Niño on the Great Barrier Reef, but nature is cruel, and there is no justification to transfer that blame to humanity.

ACADEMIC FREEDOM AND BUCKING THE TREND

Ridd's whistle blowing about poorly conducted studies by colleagues led to him being sacked by James Cook University, who said that he had brought it and associated institutions into disrepute. With the help of AUS$260,000 of crowd funding, raised in five days because of the publicity the case caused in Australia, he challenged this in court and was fully vindicated on all 28 counts, the judge expressing great concern about the lack of academic freedom at his and other universities. "To use the vernacular: the university has played the man and not the ball." He seemed to believe that universities are losing the desire to be controversial and tolerate the presentation of unconventional ideas. "Incredibly, the university has not understood the concept of intellectual freedom. In the search for truth it is an unfortunate consequence that some people may feel denigrated, offended, hurt or upset. It may not always be possible to act collegiately when two diametrically opposed views clash in the search for truth… That is why intellectual freedom is so important. It allows academics to express their opinions without fear of reprisal.

"It allows a Charles Darwin to break free of the constraints of creationism. It allows an Albert Einstein to break free of the constraints of Newtonian physics. It allows the human race to question conventional wisdom in the neverending search for knowledge and truth, and that at its core is what higher learning is about. To suggest otherwise is to ignore why universities were created and why critically focused academics remain central to all that university teaching claims to offer."

Ridd believes that only scientists nearing retirement like him dare speak out about flawed science in academia because vice chancellors and accountants place internal codes of conduct over academic freedom in the competitive field of research funding. The professors, who used to ensure academic freedom, now have little power. The university appealed against the award of AUS$1.2 million in damages and penalties to Ridd, using public money and a barrister reputed to charge AUS$20,000 per day. Due to the level of media interest, his target of AUS$1.5 millon in crowd funding to defend himself was again rapidly reached. The university won its appeal to the judges on a two-to-one decision.

In its judgement, the Federal Court seemed to be realigning what is expected of freedom of speech for the internet era. "There is little to be gained in resorting to historical concepts and definitions of academic freedom. Whatever the concept once meant, it has evolved to take into account contemporary circumstances which present a challenge to it... Academic freedom plays an indispensable role in fulfilling the mission of the university... But a host of new challenges have arisen in recent years in response to the changing norms and expectations of the university. With the increasing role of the internet in research, the rise of social media in both professional and extramural exchanges, and student demands for accommodations such as content warnings and **safe spaces**, the parameters of, and challenges to, academic freedom often leave us in uncharted territory." "Safe spaces" from harsh scientific reality?

This is a story of how climate change orthodoxy has become so engrained that to challenge it is to risk one's reputation and career, like Bill Gray, the foremost hurricane expert in America in the 1990s who told Al Gore that doubling CO_2 levels would produce less than a 0.5°C temperature rise. His public funding stopped immediately. Gray was reputed to be a blunt free thinker who often railed against the establishment, but we need people like him to keep others on their toes.

Those who run academic institutions have a difficult job. They are accountable for public money, so must take note of political policies and sensitivities, yet retain enough independence to follow a robust academic agenda that should be non-political. They must remain open to internal academic controversy, yet retain a reputation for stability and prudence to attract students and research funding. My own experience is of two years on the board of a large further education college. Board members like me were charged, on behalf of the public, with ensuring that college policies and practice were appropriate and competent. Board meetings were behind closed doors, where we could be as critical as we thought appropriate, but the minutes were made public so were 'sanitised' such that we hardly recognised them as representing the meeting we had attended. We hoped that the college staff who attended the meeting had remembered what was said and took appropriate action. If politicians and the media had read an honest account of proceedings they could have sensationalised them, which might have tarnished the good reputation of the college.

The case went to the High Court of Australia in February 2021, and the judgement in October was against Ridd. He had infringed his contract of employment by not restricting his criticisms to his area of expertise, that is oceanography, but extending it to the university's quality control mechanisms, and in making his comments public, even sharing them with his wife being forbidden. In response, Ridd wrote: "Secrecy, together with spying on communications, is an incredibly powerful tool in the hands of any authoritarian organisation, whether it be the Stasi or an Australian University. It silences dissent, isolates the victim and establishes where the power lies – with the university. The academic is crushed." In addition to the AUS$1.5 million in public contributions, Ridd lost AUS$300,000 of his own money but, despite losing the case, the judges did not ask him to pay the university's legal costs.

The publicity surrounding the case led to Australian Education Minister Alan Tudge introducing new legislation on academic freedom, recognising that "There are few things more important for the advancement of truth and knowledge than having open, robust debate at our universities", but it would not have changed this legal decision. I have attended conferences which were held under Chatham House Rules, whose principle is: share the information you receive, but do not reveal the identity of who said it. Politicians could speak openly and honestly without the press attributing comments to them. Democracy relies on free speech but pragmatism often precludes it, a dilemma that we struggle to reconcile, with politicians invariably accused of insincerity and hypocrisy, as if they cannot be excused human frailties. Consider Peter Ridd's dilemma. If he did not make public the details of his concerns, he could not obtain the crowd funding to fight his legal case against the university, which had large public funds at its disposal. Should he have kept quiet when Queensland farmers were accused of allowing fertiliser and herbicide run-off to harm the Great Barrier Reef when he had convincing evidence that the reef was not being damaged? Could he have been a concerned member of the public and an academic at the same time?

Discussion: If publicly funded institutions are wary about washing their dirty linen in public because of reputational damage, does this inevitably compromise open public accountability? Should there be 'safe spaces' from controversy and the search for truth at universities? If trolling is a problem, should the response be to restrict any views and comments which someone might conceivably find offensive, even if they are supported by verifiable data?

QUALITY CONTROL AND PEER REVIEW

The judge in the Peter Ridd Federal Circuit Court case which he won cited the independent thought of Darwin and Einstein, but they did not work within the field of publicly funded academic research, a largely post-1950 phenomenon. Incidentally, 200 fellow scientists initially poured scorn on Einstein's theories, to which he responded that he did not need 200 critics to prove him wrong, only one with verifiable data. Scientific research is not a popularity contest, so when you hear that "97% of scientists agree that..." be sure to check the terms of the survey and the question that was actually asked. It is credible to assert that 97% of climate scientists reckon that human emissions play some part in climate change, but there is a wide range of opinions on how much this is. Only one will be proved correct, and it will be credible data that will do so, not a straw poll.

Ridd's main crime was to criticise the university's quality control procedures, which allowed unverified and inaccurate studies to be published. Quality control in academic research is largely done by fellow researchers reviewing a study's findings. This is more about checking for process than content, and is often done by sympathetic people in the same line of research, such that critics refer to it as 'pal review' rather than peer review. Very few academic research projects are subject to replication studies before their findings are published in journals, and the summary of their findings reported in newspapers, if they are deemed dramatic enough by editors to appeal to the public.

This is a serious problem, and not just in climate science. In 2016 *Nature* magazine published a study revealing that, "more than 70% of researchers have tried and failed to reproduce another scientist's experiments, and more than half have failed to reproduce their own experiments". The three main weaknesses were seen to be selective reporting, pressure to publish and poor analysis. Cancer researchers C.G. Begley and Lee Ellis tested 53 landmark studies and could only replicate six (11%). In 1998 Fiona Godlee, editor of the *British Medical Journal*, sent an article containing eight deliberate mistakes to more than 200 of the BMJ's regular reviewers. Not one picked out all the mistakes. On average, they reported fewer than two; some did not spot any.

Richard Horton, editor-in-chief of *The Lancet*, in *The Times* of 4 March 2016 wrote bleakly: "The case against science is straightforward: much of the scientific literature, perhaps half, may simply be untrue." In 2000 he wrote: "We portray peer review to the public as a quasi-sacred process that helps to make science our most objective truth teller. But we know that the system of peer review is biased, unjust, unaccountable, incomplete, easily fixed, often insulting, usually ignorant, occasionally foolish, and frequently wrong."

Writing in 2006, then 87 years old, James Lovelock (he died in 2022 at the age of 103) despaired of the state of academic research and the power of state institutions to stifle initiative. "Younger scientists cannot freely express their opinions without risking their ability to apply for grants or publish papers. Much worse than this, few of them can now follow that strange and serendipitous path that leads to deep discovery. They are not constrained by political or theological tyrannies, but by the ever-clinging hands of the jobsworths that form the vast tribe of the qualified but hampering middle management and the safety officials that surround them."

Depressingly, the media promotes bad science, misinformation and alarm. Here is a quote from the TV programme *Blue Planet Revisited: Great Barrier Reef*. "If high temperatures are maintained, corals bleached in this way are likely to die. Since 2016 half of the Great Barrier Reef's shallow corals have perished due to bleaching. If we don't change our habits with regards to CO_2 emissions I do not believe the coral reefs will be here in the future. On Heron Island, Professor Sophie Dove (the University of Queensland) is on a race against time, trying to predict what the future may look like for coral reefs. By creating miniature reefs in a set of tanks she can find out how corals might react to future climate conditions. The aim is to reproduce **a structure that looks a little bit at least what we have out there on the reef slope** so that we can then examine what increases in temperature and acidification do to the mini reefs. Sophie's experiments have shown that if temperatures continue to rise at their current rate we will lose our coral reefs in just thirty years."

Dove is creating an artificial laboratory experiment which does not replicate realistic conditions ("looks a little bit at least what we have out there") and which may produce any desired result. Building on a grossly false premise ("half of the Great Barrier Reef's shallow corals have perished"), irresponsible, speculative predictions are made ("we will lose our coral reefs in just 30 years"). If this documentary had been made by an extremist group like Extinction Rebellion or Greenpeace this would have been bad enough, but it was produced by the publicly funded BBC.

The truth, as always, is not in the drama promoted by activists, but in the data, which has been collected by the Australian Institute of Marine Science for almost four decades. It shows that the average reef cover over the whole length of the GBR has generally been around 15-20% of the sea floor in the study area. This dropped to 10% in 2012 but was over 30% in 2021. No one can explain these variations but the reef is currently in the best condition it has been since records began. It is doing just fine. It seems that this is another example, like climate change, where we fill a gap in our understanding of nature with demons and delusions of our own making.

Ocean acidification is a myth promoted by undemocratic, unaccountable, enigmatic, patrician organisations like the UN and the BBC, which have similar didactic mindsets. Peter Ridd is correct. There is nothing that humans are doing which is significantly harming the vast ecosystems that comprise the GBR, or perhaps could possibly do. We are neither the master nor the destroyer of the planet and its natural systems, just an irritator of small parts of it. We should ignore the propaganda, observe the verifiable data, and avoid unnecessary anxiety.

IS MORE CO_2 IN THE OCEANS GOOD OR BAD?

Most scientists have been looking for the downside of more CO_2 in the oceans, but *Ow et al.* 2015 found that elevated CO_2 levels caused enhanced photosynthetic responses in seagrasses, just as more CO_2 in the atmosphere increases plant growth on land. That can only be good news for the dugongs and other creatures that feed on them and live in them. I made the point earlier that phytoplankton, microscopically small plant life, must benefit from more fertilising CO_2. *Miller et al.* 2013 found that higher CO_2 levels stimulated the breeding activity of clownfish in the laboratory, and in the fringing reefs of Orpheus Island in the Great Barrier Reef clutches per breeding pair produced

67% more eggs than the control sample, although the quality of eggs was poorer. Different species respond to very high levels of CO_2 (up to 2,850 ppm) in different ways in the laboratory, with oysters, clams, scallops and conchs slowing down the rate of shell building, but crabs and lobsters speeding it up (*Ries et al.* 2009).

Activists would have us believe that varying ocean acidity/alkalinity can only be harmful, but they underestimate the resilience that corals, and the algae that live on them, have to respond to change. Once again, theoretical laboratory experiments and computer models, which cannot take on board the complexity of natural systems, are substituted for robust scientific research.

The next chapter will deal with threats to the environment due to human actions. Environmentalists react with horror to waste plastic making its way into the oceans, but a 2021 study in *Nature Communications* by the Smithsonian Environmental Research Center found that plastic debris in the Pacific, including microplastic fragments less than 5 mm wide, was being colonised by sea creatures, just as they colonise shipwrecks. Crabs, Asian anemonies, marine worms, sea stars and sponges were found far from their normal habitats, where the open ocean had previously been a marine desert. Author Greg Ruiz was surprised by this, not only because ocean and coastal life was thriving among plastic debris, but he could not understand where the food is coming from in mid-ocean. Perhaps it is more phytoplankton due to more CO_2 in the water.

I am not arguing for more plastic debris to be dumped in the oceans, but it seems inevitable that people in developing countries who cannot afford good waste treatment facilities, and have the poorest in society rummaging through dustbins and waste dumps, will allow debris of all types, including plastics, to escape into the environment. Once again it is assumed that all human actions will be detrimental, but perhaps there are benefits for organisms adaptable enough to survive and even prosper from our excess and largess.

Discussion: Some scientists have placed sea creatures with calcium shells in water that has been artificially made more acidic and noted the consequences, often with harmful and alarming results. To be more realistic, the water alkaline level must be reduced slowly over many days by percolating CO_2 through it. Do you think that better standards should be set for research scientists to adhere to, and can you imagine a better way of checking experimental results? Who should set these standards? Who should pay for replication studies (essentially rerunning the experiments independently) which do not happen just now? Should we ban the publication of experiment results in the press (as opposed to purely academic journals) until replication studies have confirmed their veracity?

CHAPTER 6
WHAT ABOUT SPECIES EXTINCTION?
ECOLOGY LESSON

I watched several speeches and interviews with Greta Thunberg when she visited America in 2019. Almost everything she said was preaching about the danger of climate change and how adults were not taking the problem seriously. She only mentioned one 'fact': that species were going extinct at a rate of up to 200 per day. She did not say where this figure came from, so we cannot check it.

Since the late 1960s, the International Union for the Conservation of Nature (IUCN) has been reviewing species which may be threatened. This is a positive process because it alerts us to where conservation efforts should be directed. In the year 2000, about 16,000 species had been assessed and 11,000 were judged to be threatened, a high percentage but it obviously chose to review first the ones which were believed to be in danger, and it still does to a certain extent. In 2016 it reported about 24,000 species that were threatened out of a total assessed of over 80,000. If we were to divide 24,000 by 200, all threatened species would be gone in 120 days, or about three months, that is before the end of 2019. I have checked. This did not happen. Using the IUCN information, the figure is less than two species going extinct per year over the last few centuries, and during most of this time human emissions were small.

CLIMATE CHANGE IS NOT THE PRIME THREAT TO THE NATURAL ENVIRONMENT

The IUCN does not place climate change as one of the top threats to the natural environment. These are: overexploitation (like overfishing), agriculture (land taken away from nature and poor use of pesticides and herbicides), urbanisation (loss of habitat to buildings and roads), invasive species and disease (like American grey squirrels pushing out native British red squirrels), pollution (like human waste products fed into rivers and the sea) and modification of ecosystems (like large areas of forest removed for agriculture). These are issues we know how to solve, and they come way above any problems that climate change presents, like a 200–300 mm rise in ocean level over a century, or a half degree centigrade rise in temperature over 40 years. We can only speculate about what the climate might do in the future, but it seems at the moment that climate change is not a big factor in species extinction.

On 14 June 2016, *The Guardian* announced: "the first recorded mammalian extinction due to anthropogenic climate change". The poor Melomys was the only endemic mammal species on the Great Barrier Reef and it lived on a single island. It was last seen in 2009 and a search for it in 2014 proved fruitless. Apparently, at high tide the area of the island dropped from 4 ha in 1998 to 2.5 ha in 2014, so rising sea levels and therefore global warming must be to blame by reducing its habitat.

The fact that *The Guardian* was only reporting this two years later suggests that it fitted with the general doom and gloom it was then covering on the reef (which was being damaged by the 2015/16 large El Niño event, not long term climate change). The report does not explain how this creature got to the island in the first place – surely it could not have evolved there? There must have been lots of these creatures on the mainland and other islands which have become extinct for the normal reasons – natural evolution (by far the biggest cause of extinction), invasive species and loss of habitat, the last two possibly due to humankind's activities. Or perhaps they have just been overlooked or mistaken for another species of rat. We should look out for news reports saying that the Melomys has been found elsewhere. The Aldabra Atoll banded snail, despite being declared extinct, was later found on another island.

Some environmentalists claim that we have entered the 6th Mass Extinction of life on Earth. It is alleged that a 2°C rise in world temperatures from pre-industrial levels would cause 20% of species to go extinct. Terry Root, Senior Fellow at Stanford University Woods Institute predicted a 4°C rise in temperature and half to three-quarters of species becoming extinct in the next 200 to 300 years. How can she possibly see so far into the future, even if her assumptions were correct? She reckons that the background extinction rate is 1 in 1,000 species per 1,000 years, but that we are now on a course of 100 in 1,000 species per 1,000 years. Warming pushes species ever higher up mountains to cool off until there is nowhere else to go. More droughts remove habitats, except they might if there were more of them which, as we have seen before, there are not.

The WWF *Living Planet Report 2020* highlights the extremely sobering biodiversity issues facing us, like the 83% decline in wild animal populations in Latin America, but I have two criticisms, where the temptation to spread alarm seems to have been overpowering. Despite noting that "the IUCN Red List represents the most comprehensive and objective system for assessing the relative risk of extinction of species", it quotes the 2019 report by the Intergovernmental Science-Policy Platform on Biodiversity and Ecosystem Services (IPBES): "destruction of ecosystems has led to 1 million species (500,000 animals and 500,000 insects) being threatened with extinction over the coming decades to centuries". In contrast, of the 100,000 or so species it has now surveyed, the IUCN calculates that 6% of species are critically endangered, 9% are endangered and 12% are vulnerable to becoming endangered.

We cannot just statistically apply these percentages to estimates of possible numbers of species we have not yet studied in a 'species area model' (yet another artificial model of the real world), or treat vulnerable species as seemingly critically endangered if we extend the timescale out over centuries and ignore the steps we are already taking, and will increasingly take, to address the problems. Also, instead of concentrating on how humans have influenced extinctions since around 1850, when industrialisation and population growth took off and when we have better records, IPBES uses the year 1500 as the start of the timeline: "Humans have driven at least 680 species of vertibrates, the best studied taxonomic group, to extinction since 1500." We know how human populations suffered due to the Little Ice Age up to the beginning of the 19th century, so surely more vulnerable wild populations must have been under greater threat from the extreme cold?

The second criticism is the focus on climate change despite the WWF report saying: "climate change has not been the most important driver of the loss of biodiversity to date, yet in coming decades it is projected to become as, or more, important than the other drivers". We have enough to sort out based on the scientific data we have already collected, and the factors we have already observed, without speculative projections about future ones.

POLAR BEARS AS THE ICON OF GLOBAL WARMING CAMPAIGNERS

When Greta Thunberg was a young girl she probably saw a documentary by David Attenborough showing polar bears in distress, with the conclusion that melting Arctic sea ice was causing them to starve. A few years earlier Al Gore, in his film *An Inconvenient Truth*, had made similar allegations. This was very scary, because it seemed to show climate change in action, in the cruellest of ways. Even in 2019, the polar bear was seen as an icon of harmful climate change, as witnessed by a backdrop photo in a discussion of the subject between David Attenborough and Prince William on TV.

When I was a young adult, around 1970, I saw a documentary which showed fishermen clubbing baby seals to death in an attempt to stop them eating the fish that the men wished to catch. The red blood on the white snow was shocking. There was such a public outcry that conservation measures were introduced in 1973 by all the countries bordering the Arctic. The baby seals still suffered a very cruel death, but from being mauled by polar bears, whose numbers rose from about 5,000 to almost 30,000 now. Overpopulation of polar bears is causing a problem in human settlements in the Arctic but, as custodians of nature, it is sensible to deal sensitively with the situation, which does not involve shooting them indiscriminately. Of course, activists claim that it is melting ice that is making the polar bears come out of the water and bother humans.

Susan Crockford (University of Victoria to 2019): "The US Geological Survey estimated the global population of polar bears at 24,500 in 2005. In 2015, the IUCN Polar Bear Specialist Group (PBSG) estimated the population at 26,000 (range 22,000–31,000) but additional surveys published since then bring the total to near 28,500 with a relatively wide margin of error." In 1996, the IUCN considered moving polar bears from "endangered" to the "lower risk, conservation dependent" category, but was persuaded that climate change may mean that the population will decline later in the 21st century. In 2007 Christine Hunter and colleagues from the US Geological Survey in Virginia stated that predicted summer sea ice decline by 2050 would cause a 67% decline in numbers, as if such speculation could be described as scientific. This was the first time that the IUCN had justified a species as being endangered, not on actual evidence of habitat loss, but of imagined future loss. The polar bear is a great conservation success story, but it is not recognised as such by those who should know better.

CLIMATE CHANGE AS THE DEFAULT REASON FOR ENVIRONMENTAL PROBLEMS

The bias of nature documentary programmes has been evident in recent years where the voice of Attenborough has been used to infer unimpeachable authority. Climate change is mentioned at every opportunity, although very few of the examples of its effects are properly justified. In the Netflix 2019 programme *Our Planet*, walruses were filmed plunging to their deaths over cliffs with the contention that shrinking Arctic sea ice, caused by AGW, had unnaturally forced them on to land where, in their desperation, they climbed steep slopes and fell to their deaths. Episode 2 of BBC's 2019 *Seven Continents, One Planet* also featured this but added polar bears, also allegedly forced on to land by shrinking ice, to scare walruses over the precipice. In Episode 1, Attenborough claimed that warming Antarctic waters caused more fractured ice which made (cute) penguins more susceptible to sea leopard attacks, and increased storms in southern oceans blew poor little albatross chicks off their nests to their deaths.

These glib, throw-away comments do not stand up to scrutiny when real scientific enquiry is applied. There is no evidence that southern oceans have been warming since records began in the 1980s, that there is more fractured sea ice, or that storms have been increasing in intensity; quite the contrary. According to *IPCC AR5*, "it is *very likely* that the annual mean Antarctic sea ice extent increased at a rate in the range of 1.2% to 1.8% per decade between 1979 and 2012". The evolutionary process will decide whether albatross' instincts that make them construct their nests raised up from the ground, and have adults which cannot recognise their young when they are beside their nests and not in them, are good survival strategies.

Climate change, whatever the actuality or cause, is irrelevant. What human beings can do (*The Guardian* 31 January 2019) is police sustainable fishing regulations, for example 'night-setting' where the baited long fishing lines are laid at night when albatrosses and petrels don't feed, using heavier lines that sink below the surface quicker, and using streamer lines to scare birds, to reduce the estimated 100,000 birds hooked and drowned each year. Fifteen out of twenty-two albatross species are on the IUCN endangered list and, whatever part the rest of nature is playing in this, we can make sure that we do our bit.

A three-year study by the US Fish and Wildlife Service, starting in 1994 at Cape Pierce in Alaska, identified the suicidal trait in walrus behavior that Attenborough's programmes featured. They noted the habit of them resting on land to bask in large numbers in the few months of summer sunshine but, apart from speculating that overpopulation might be a contributory factor, could determine no reason for the habit of climbing hills that they had no ability to safely descend. But a few hundred deaths out of 12,500 basking seals over three years does not threaten the sustainability of the species. Polar bears are taking advantage of free meals and their numbers are also rising. This is all an ecological good news story, but Attenborough and colleagues have turned it into a people-entertainment horror story. The 1994 scientists in Alaska admitted their ignorance and the possibility that human logic does not work when evolutionary processes are unfolding: "The best approach to this is let nature take its course."

Despite his pessimism, Attenborough found the sea life in good condition and concluded: "These seas are once again brimming with life and scientists have discovered that the Southern Ocean and the life within it soak up more than twice as much carbon from the atmosphere as the Amazon rainforest. By protecting the Antarctic we don't just protect the life here, we are helping to restore the natural balance of the entire planet." He could have put it another way: "Thanks to our CO_2 emissions, the cool Southern Ocean is absorbing more of this gas, making it more fertile for sea plants and creatures, and so our actions have caused the waters to brim with life." But this would be just as mischievous without solid scientific data to support it.

When fact checking is short-circuited, conclusions depend purely on our frame of mind at the time (see Chapter 11 for the 'Garden of Eden' complex – the world was a perfect place before humans defiled it). Protecting wildlife where we can is simply a good idea. "Restoring the natural balance of the entire planet", reducing the world's human population (Attenborough's Population Matters organisation), or controlling the whole world's climate with a CO_2 dimmer switch, are just fanciful. Only the archetypal 'mad scientist' would have such unrealistic, megalomaniac ambitions.

Discussion: When authors make statements in books, or reporters refer to scientific studies in newspapers,

they usually give the source of the information so that the reader can check out the facts for themselves. This does not happen on television programmes, where emotive pictures of suffering wildlife or destructive storms (with dramatic musical accompaniment) can be used to infer that individual events are representative of a larger problem, despite there being no long-term trend in statistics. This is bad science. How could this be addressed?

THE ANALYSIS OF ACTIVISTS IS SKEWED AND UNBALANCED

People like Attenborough and Gore do not mention that wind farms chop up birds and bats (YouTube *Wind Turbines v Birds and Bats*), and solar farms fry birds which confuse them for lakes. This may not be in large numbers (although we don't really know), but the populations of the raptors and seabirds they kill are smaller and more sensitive to threats than, for example, to garden birds. The RSPB is concerned that puffin and gannet colonies are threatened by massive offshore wind farms in Scotland where a third of Europe's seabirds breed. We don't know how big the problem is because the body parts are washed away in the sea, and on land they are quickly eaten by small carnivores.

Our response to a problem (climate change) involves a cure with unintended harmful consequences. The climate activists must see such casualties as unfortunate but necessary 'collateral damage' in the battle to save the planet, whereas if dozens of dead birds were found beside any other industrial facility, they would be up in arms. We also ignore the fact that, as Matt Ridley (*The Rational Optimist* and *The Evolution of Everything*) points out, 700,000 ha of forest in SE Asia has been cut down to produce palm oil for bio-diesel in Europe, and that 5% of the world's grain crop is used for car fuel in areas that might better be reforested.

Bjorn Lomborg was a member of Greenpeace but decided to analyse its statistics, and those of other organisations like WWF. It was claimed that 250,000 seabirds were killed in the Exxon Valdez oil spill in Prince William Sound in 1989. In his book *The Skeptical Environmentalist* he noted that this was the number of birds which were killed flying into plate glass windows in the US every day, or the number of birds killed by domestic cats in the UK every two days. Two billion dollars was spent on the oil spill clean-up campaign but it was later found that beaches which had been cleaned took four years to recover while ones which had been left alone to clean naturally took eighteen months to recover. As James Lovelock (*The Revenge of Gaia* 2006) said: "When an accident at sea releases huge volumes of crude oil onto beaches, rocks and coves we see it as an environmental disaster, and not long ago we tried vainly to wash it away with detergents. Now, with greater common sense, we leave the clean-up to the natural organisms that regard the spillage as food."

Activism and common sense do not seem to go together. What did make sense was increasing regulations to make oil spills less likely (there is also a financial incentive for the oil companies not to lose oil and bare the consequences of spills), such that major spills have dropped from an average of 25 per year in the 1970s to less than two per year in the 2010s.

GOOD, BALANCED ANALYSIS LEADS TO APPROPRIATE REMEDIES

Matt Ridley considers the IUCN figures from a more positive standpoint, based on what has happened rather than what might happen. "Over the past 500 years, we know of 77 mammal species (out of about 5,000) and 140 bird species (out of about 10,000) that have gone totally extinct. There may be a handful more we do not know about, and there are plenty more on the brink. Nonetheless, these are the official total species extinctions for the two groups of animal we know best, as compiled by the International Union for Conservation of Nature. Of those 217 species of bird and mammal, almost all lived on islands – if you count Australia as an island – and just nine on continents." We have proved we can do something about that, and we spend at least $21.5 billion every year on global biodiversity conservation (*IFL Science* 21 March 2016). Two hundred and thirty-six native species on 181 islands have been protected by eradicating invasive species.

Ridley: "My point is not to say extinction does not matter, but to try to get at the real cause of the extinction surge, and it is clearly not the growth of human population, which has mostly happened on continents. Europe has lost just one breeding bird in 500 years, for example – the island-breeding great auk in 1844. It is true that there was a surge in extinctions of mammals in the American continent about 12,000 years ago, but that was caused by hunter-gatherers with stone-tipped spears, not modern people with cars. Indeed, there is a pretty spectacular revival of wildlife today in rich continents like North America and Europe. Modern prosperity is plainly not the cause of animal extinctions." If those who see economic growth as the problem, rather than the solution, were to get their way, improvements in conservation would be impaired.

"The cause is man-made, all right, but it's not because we killed them or destroyed their habitat. It's the rats, cats, goats, pigs, mosquitoes and avian malaria we brought with us that did the damage on Hawaii and throughout the Caribbean, the south Atlantic, the Indian Ocean and the rest of the Pacific. The dodo disappeared from Mauritius not because sailors ate them (though they did) but because of predation by monkeys, pigs, rats and the like… Closer to home, it's invasive species that are the main cause of conservation problems: grey squirrels, mink and signal crayfish. Why are wolves increasing, lions decreasing and tigers now holding their own? Because wolves live in rich countries, lions in poor countries and tigers in middle-income countries." Polar bears also live in rich countries and, as we have seen, they are thriving and not in danger of extinction from climate change or any other cause.

"Misdiagnosing the cause of extinction leads to mistaken policies… It is not just dishonest to pretend that the way to prevent species extinctions in the future is to throttle back on economic activity and to spend a fortune fighting climate change. It is also dangerous. It has already slowed and hindered the conservation movement from focussing on the real, growing and urgent threat to native wildlife all over the world, and especially on islands. We have to get serious about bio-security, about regulating the trade in live animals and plants, which results in more and more alien species jumping ecosystems where they run amok.

"In Europe and America, rivers, lakes, seas and the air are getting cleaner all the time. The Thames has less sewage and more fish. Lake Erie's water snakes, on the brink of extinction in the 1960s, are now abundant. Bald eagles have boomed. Swedish birds' eggs have 75% fewer pollutants in them than in the 1960s... Now that weeds can be controlled by herbicides rather than ploughing (the main function of a plough is to bury weeds), more and more crops are sown directly into the ground without tilling. This reduces soil erosion, silt run-off and the massacre of small animals of the soil... The grass meadow near my house, sprinkled with nitrate twice a year, supports a large herd of milking cows, but it is also teeming with worms, leatherjackets, dung flies – and the blackbirds, jackdaws and swallows that eat them."

Instead of looking forward with speculative pessimism, we can look back on our success stories so far as custodians of nature. Callum Roberts and colleagues at the University of York published a report in *Nature* in April 2020. It had lots of good news. Of the 124 populations of marine animals studied, 47% had increased, 40% were stable and only 13% were decreasing. Marine Protected Areas have increased from 0.9% of the ocean in 2000 to 7.4% in 2019, which is remarkable progress. Humpback whales migrating from Antarctica to Australia have increased from a few hundred in 1968 to more than 40,000 now. Gray whales have increased from 4,000 in 1949 to 20,000, and blue whales from about 1,000 to 10–20,000 today. Northern elephant seals, almost extinct by 1880, now number more than 200,000. There were about 50 southern sea otters in Canada in 1911 but several thousand now. A turtle conservation scheme on Chichijima island in Japan increased breeding females from 50 in the 1970s to about 600 in 2010.

I am a (relatively) rich person in a rich country. I can afford to purchase a house at the edge of town with a garden that is 18 metres squared. It allows me to grow flowers for pleasure and fruit and vegetables to eat, while providing a habitat that attracts wildlife. About 20 species of bird visit my feeding stations and a small pond which has breeding populations of frogs, damsel flies, pond skaters and snails. I am careful, but not obsessively so, about the chemicals I use in my garden. A pair of mice overwinter in my compost heap where they can feast on worms and grubs (sorry to disturb you each spring). A bravely hungry deer entered my garden to eat berries on a shrub one winter when I left the gate open, although I usually keep my gate shut to deter foxes. I watch a wasp scrape wood from my garden fence to build its nest, where I am helping it, and in return it kills aphids that might attack my flowers. My local council used to pipe rainwater straight into underground pipes, but there is now a SUDS (Sustainable Urban Drainage Scheme) a few hundred metres from my house which collects the water in balancing ponds to provide a habitat for otters, kingfishers and newts. People invariably love their environment, and rich people like me do what they can to preserve and enjoy it.

HUMAN PROGRESS AND HUSBANDRY CAN BE THE SOLUTION, NOT THE PROBLEM

In 1851, American Herman Melville published his book *Moby Dick* about the obsessive quest of Captain Ahab to kill a giant sperm whale. Melville had spent four years as a sailor on whaling vessels, this industry being very important because whale oil had superseded candles as the best way to fuel lamps. Many species of whale were heading for extinction until George Bissell's discovery of oil in Pennsylvania in 1853 as an alternative. Gas lighting replaced oil lamps before electricity superseded both, making human lives easier and safer, and also causing less harm to the environment.

The whaling industry declined from the 1870s but some countries still hunted whales for meat until bans were put in place from the late 1980s. In theory, only 'aboriginal whaling' now takes place (killing on a subsistence basis for communities where this practice pre-dated modern times), but this still leads to around 20,000 whales being killed each year, chiefly in Canada, Greenland, the Faroe Islands, Norway and Japan. This is thought by these countries to be sustainable.

We are also better at managing fish stocks. The UK Marine Conservation Society produces a good fish guide to highlight which fish species are in plentiful supply and which are being over-caught. A system of fish quotas has been in place for many decades which has been accepted by fishermen because their interest is in conserving fish stocks for the future. For example, in 2015 the EU banned commercial fishing of sea bass in the spawning season in February and March. In 2019 the Society deemed the sustainable choices to be oysters, mussels, king prawns (British farmed), Atlantic halibut, North Sea herring, North Sea plaice and European hake. We can eat as much of these as we wish (at present). We should avoid North Sea British cod, European eel, wild Atlantic salmon, West Coast British whiting and wild Atlantic halibut.

Michael Shellenberger (*Apocalypse Never: Why Environmental Alarmism Hurts Us All* 2020): "In late 2015, the US Food and Drug Administration approved a genetically modified salmon, one that delivered major environmental benefits over existing farmed salmon. Critics loved it, 'The flesh is exquisite,' wrote one food writer. 'Buttery, light, juicy. Just as Atlantic salmon should be.' The AquAdvantage salmon, developed by AquaBounty Technologies in 1989, grows twice as fast and needs 20% less feed than Atlantic salmon. While eight pounds of feed is needed to harvest one pound of beef, only one pound of feed is required for one pound of AquAdvantage salmon.

"Unlike the majority of farmed salmon, which is produced in floating cages in coastal areas, AquAdvantage is produced in hatcheries and facilities in warehouses on land. It thus minimises the impact of aquaculture on natural ocean environments and prevents harmful infections with wild species… By genetically altering the salmon, AquaBounty also eliminated the need for antibiotics which public health officials warn can contribute to antibiotic resistance.

"Today, 90% of the world's fish stocks are either overfished or at capacity, meaning they are close to or just barely above the maximum they can be harvested before seeing their populations collapse entirely... The good news is that fish farming, or aquaculture, is developing rapidly. Aquaculture output doubled between 2000 and 2014, and today it produces half of all fish for human consumption [The State of World Fisheries and Aquaculture 2016]... And yet, the most outspoken critics of replacing wild fish consumption with farmed fish are environmental groups... In response, several large supermarket chains, including Trader Joe's and Whole Foods, announced they would not carry the AquaBounty fish, even though spokespersons from both chains admitted that the stores offer other foods produced with genetically modified ingredients or feed."

Callum Roberts's team estimated that continuation of world conservation measures, including a target of 30% of the ocean being protected by 2030, would cost between $20 billion and $40 billion per year. The higher figure would be $5 per person on the planet which is easily within our means (or $30 per person per year if only richer countries contributed). That seems excellent value for money and eminently good common sense. While the WWF's speculation about the future is questionable because practically all predictions are wrong, it attempts to be reasonable, adopting the IPCC 'middle-of-the-road' scenario with population peaking at 9.4 billion by 2070 and assuming continued economic growth. Its "half-Earth" target of 50% of the planet being protected for biodiversity conservation is probably achievable given the progress we have made so far to the current 15% (plus another 25% that is currently 'wilderness'), and it does include genetic modification in the mix for "sustainable intensification of agriculture production". But it also seems to have to include highly inefficient 'organic farming' for the part of its membership which religiously supports this. Its *Living Planet*

Report 2020 is, after all, a marketing document for prospective new members and for retaining the 5 million existing ones who support it financially.

Panicking about climate change, and wasting resources on many initiatives which might make us feel better about ourselves but make no significant contribution to lower human emissions (see Chapter 10), is foolish. Some of these initiatives, like cutting down forests in Malaysia and Indonesia to produce palm oil for biofuels, are actually causing huge damage to the environment, and destroying the habitat of many animals.

Discussion: Human beings have caused the extinction of some species like the dodo. Would it be a bad thing to completely eradicate species of mosquito that cause malaria and some other harmful diseases, the way that science eradicated the disease smallpox by 1980? Should all natural organisms be regarded as precious?

It seems to me, in the many nature programmes I have watched on TV, that the fight for survival in the wild, due to the constant competition for food and the vagaries of the weather, is extraordinarily fierce, with populations waxing and waning dramatically over years and decades. The natural world is a match for any harm that humans can do, so we should not beat ourselves up too much. Genetic research suggests that the population of Homo sapiens may have dropped to very low levels, almost to extinction, until it recovered to populate most parts of the planet. We are now helping other species to survive, not just by creating urban environments that are successful homes for many birds and mammals, but reaching out to rid areas of invasive species and protect selected habitats. Let's just do more of it.

SAVING THE PLANET FROM ITS OWN SELF-DESTRUCTION?
I want to end this chapter broadening our perception of the planet, to move the focus far back beyond the last two hundred

years, which is called by some the Anthropocene, the Human Epoch, as if humans have the same power to transform the Earth as previous geological and solar system forces have, and cause a mass extinction event. When we look backwards we find that the current era contains the coldest temperatures and the lowest atmospheric carbon dioxide content in the last half billion years (Figure 16), during which time the ancestors of all life forms currently on the planet were evolving. An increase of one or two degrees in temperature or even a doubling of CO_2 are small in the context of tens of millions of years.

I have used the dates for the various geological periods agreed in 2009, since there has always been debate about prehistoric timescales which can only be approximate. The Ancient Greek names given to these eras provide intellectual kudos but inadvertently deter the layperson from entering this field of human knowledge, because it sounds as if it is too clever for the likes of you and me, but Holocene, Pleistocene, Pliocene and Miocene merely mean 'extremely new', 'most new', 'more recent' and 'less recent' respectively.

The Holocene is the period since we emerged from the last Ice Age, officially 11,650 years ago. Figure 36 in Chapter 12 suggests that the Holocene temperature peaked about 9,000 years ago and the world is slowly cooling as we return to Ice Age conditions in a 100,000-year Milankovitch cycle that has been taking place throughout the Pleistocene. In other words, the Holocene is merely a small part of the 2.6 million-year Pleistocene, a period when the world has been as cold as any other time in the last half billion years, and the level of atmospheric CO_2 varied between about 180 parts per million atmospheric volume at the coldest points to about 280 ppm during brief interglacial times. The amount of CO_2 in the atmosphere during the Pleistocene has been determined by temperature, not the other way round as Al Gore might have us believe. At about 150 ppm plants shut down their photosynthesis process, so at certain times during the last 2.6 million years the planet has come close to eradicating all plant and animal life, a mass extinction event to end all mass extinctions.

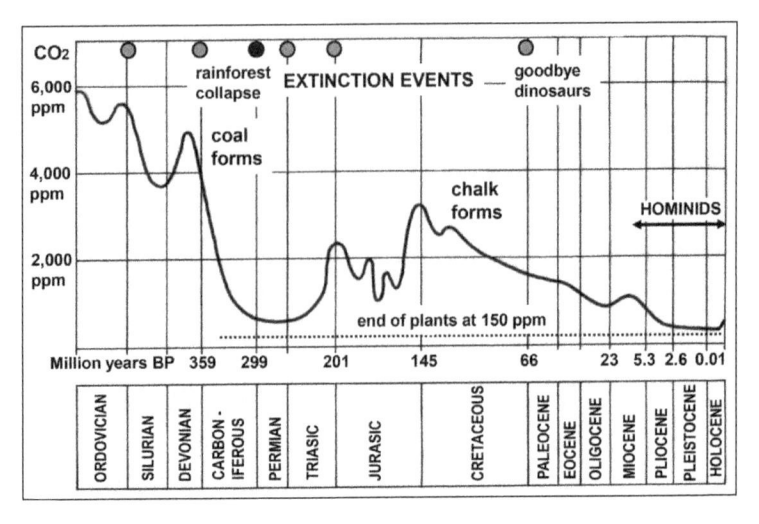

Figure 16 Historic levels of atmospheric CO_2. Source: Patrick Moore.

Prior to the Carboniferous Period, 359 million years ago, CO_2 levels were close to 5,000 ppm and the planet was perhaps 10°C warmer than today. Greater volcanic activity probably caused the high levels of CO_2. Plant life flourished and the coal we mine today was formed. Trees take in CO_2 as they grow and release most of the carbon back to the atmosphere when they die and rot, but a small percentage remains in the ground and over millions of years coal deposits formed. But in the following 60 million years both temperatures and CO_2 levels fell to roughly what we have today, with what geologists call the Carboniferous Rainforest Collapse, not quite a mass extinction event as large as some, but a big jolt to the ecosystem. It seems that the temperature and CO_2 levels fell below the level the tree species could cope with and there was a mass die-back. The temperature rose over the next 40 million years but the CO_2 level remained low, one of many examples of where the two did not correlate. The end-of-Permian mass extinction event then took place. The end-of-Ordovician mass extinction happened when the CO_2 level was over 5,000 ppm but the temperature dropped suddenly (in geological timescales) to the same levels as in our current Pleistocene ice age.

The Cretaceous era is named after the Ancient Greek word for 'chalk' because this was when chalk deposits were laid down. The continents were moving apart, the coastline increasing and the ocean currents we know today were starting to form. Sea creatures had evolved which took dissolved carbon dioxide from the sea to make calcium carbonate for their shells, which fell to the bottom of the ocean in sediments when they died. They did this so well, and thrived so much, that atmospheric CO_2 dropped from over 3,000 ppm to less than 2,000 ppm during the Cretaceous Period, and the decline kept going into the Pleistocene, dropping to the perilously low 180 ppm at some points. By exploiting fossil fuels over the last two hundred years, humans are returning some subterranean carbon to the atmosphere that was buried there mostly in the Carboniferous and Cretaceous periods, albeit we are doing so in a very small way since we will be lucky to elevate CO_2 to more than 550 ppm, very low in geological timescales. It could be said that we are making a modest contribution to stop our living planet from killing itself in another 10,000 years when a new Ice Age takes hold and the CO_2 level dips again.

Three cheers for carbon emissions! In reviewing historic data, to hypothesise possible futures, it is just as easy to make carbon dioxide the hero as the villain. Tim Lenton, in his book *Revolutions That Made the Earth*, has a diagram similar to Figure 16 showing that terrestrial creatures evolved in a carbon-rich atmosphere during the last 400 million years, with levels as high as 5,600 ppm, yet he wishes to raise alarm due to modern day emissions inevitably rising to over 500 ppm due to human progress. We need to hold such people to account. I cannot see any evidence that varying CO_2 levels caused tipping points in the climate, but I can see evidence of high densities of this gas supporting an abundance of vegetation to feed a proliferation of creatures of all sizes in the food chain, including giant dinosaurs.

CHAPTER 7
WHY POVERTY REDUCTION IS A MORE URGENT PROBLEM THAN CLIMATE CHANGE
SOCIOLOGY AND ECONOMICS LESSONS

In rich countries we tend to regard those people in less developed parts of the world, who are significantly poorer than us, as 'generally poor', without understanding the significance of more subtle levels of poverty/wealth which Hans Rosling formalised and were adopted by the United Nations. If Hans were alive today (he died in 2017) he might scold me for referring to such simplistic terms as developed and developing countries, or the West and the rest, because, while such divisions were easier to detect in the 1960s, the enormous growth in human wealth since then has led to those suffering extreme poverty going from the majority to less than 10% now. There is a spectrum of degrees of wealth in countries throughout the world. He reviewed this on various parameters such as people with big families where many children die (the great majority of the world's population in 1965) to small families and few children die (the great majority of the world's population now).

In relation to extreme poverty, Rosling said, "The suffering it causes is not unknown, and not in the future. It's a reality. It's misery, day to day, right now. It is also where Ebola outbreaks come from, because there are no health services to counter them at an early stage; and where civil wars start, because young men desperate for food and work, and with nothing to lose, tend to be more willing to join brutal guerrilla movements. It's a vicious circle: poverty leads to civil war, and civil war leads to poverty."

In the early 19th century, even in industrialising countries, average life expectancy could be as little as 30, partly due to high child mortality rates, a problem which was only solved in the second half of the 20th century with better hygiene standards, nutrition and treatment of infectious diseases. But even in poorer countries the situation is now greatly improved. "Life expectancy in low-income countries is 62 years. Most people have enough to eat, most people have access to improved water, most children are vaccinated, and most girls finish primary school... There are only a few countries in the world – exceptional places like Afghanistan or South Sudan – where fewer than 20% of girls finish primary school."

STAGES OF HUMAN DEVELOPMENT

Rosling split the 7 billion people at the beginning of the 21st century into four categories:

- Level 1: up to 1 billion people living on less than $2 income per person per day (extreme poverty).
- Level 2: about 3 billion people with an income between $2 and $8 per day each.
- Level 3: about 2 billion people with an income between $8 and $32 per day each.
- Level 4: about 1 billion people with an income more than $32 per day each.

This leads to certain typical lifestyle characteristics (see Table 1).

The wood fires used by Level 1 are not good for the forests that are being cut down, nor the children who are gradually suffocated by smoke from domestic fires, so the quicker people can afford gas and electricity in their homes the better. Electricity is also essential for job opportunities that take people away from subsistence activities and into employment. Fossil fuels are essential for people progressing through Levels 2 and 3. Some intellectuals (Level 4 ones) contend that

Table 1 Lifestyle characteristics

	Level 1	**Level 2**	**Level 3**	**Level 4**
Children per family	Five or more but high mortality	Two to five	Just over two	Two or less
Footwear	None	Plastic sandals	Robust shoes	Shoes for every occasion
Water supply	Walk to well with one plastic bucket	Cycle to well with many buckets	Plumbed water nearby or in home	Plumbed water in home (hot & cold)
Transport	Walk	Cycle	Motor cycle	Car
Cooking	Wood fire	Gas canister hob	Gas or electric hob	Gas, electric and microwave cookers
Domestic power	Wood fire	Poor electric supply for light only	Stable electricity for fridge, etc.	Reliable gas and/or electric
Food source	Meagre crops	Crops, chickens and some bought	Purchased	Purchased and eating out
Employment	Subsistence only	Low skill job	16-hour/day job when available	40-hour week normal
Healthcare	Little or none	Only if money available	Some savings to pay for treatment	Private or public insurance
Cleaning teeth	Finger	Wooden toothpick	Toothbrush	Electric toothbrush
Education	Little	Basic with house light for homework	Most children at secondary school	At least 12 years of education
Vacations	None	Too busy working for much time off	Occasional trip or vacation	Vacations, perhaps by plane

the modern world has the ability to go straight from basic technologies to modern ones, in energy supply terms missing out the intermediate coal, gas and oil and going straight to wind and solar, but as we shall see in Chapter 10, intermittent renewables and the infrastructure needed for them can only be afforded by Level 4, and they inevitably need fossil fuel backup (with further subsidies). It is not relevant to compare energy systems with telecommunications gadgets where even Level 2 can now have some access to mass-produced $20 mobile phones.

For the last 20 years on regular trips to South Africa I have watched the improvements being made to the shanty towns on the Cape Flats. Once a basic electricity and water supply has been provided to each home (tin shacks in many cases), it is possible to install solar water heating, not expensive photovoltaic panels, but water pumped through plastic tubes that are exposed to sunlight on the roof. The most notable feature has been the sprouting of rooftop satellite dishes to pick up the English Premier League (as well as local soaps), where Level 2 and 3 people can no doubt dream about rags to riches stories of poor black African footballers becoming millionaires. TVs in shacks really is advancement into the 21st century, but the electricity is predominantly powered by cheap coal, with some nuclear (I know because I specified the facilities that were required at a new hospital for any radiation accident at the single African reactor).

Almost all people reading this book only experience relative poverty, rather than extreme poverty. In America, anyone below Level 4 will be regarded as poor. In Brazil most people live at Level 3, but with many still in Level 2, about 10% at Level 4 and a small number still at Level 1. Rosling. "Most of your firsthand experiences are from Level 4; and your secondhand experiences are filtered through the mass media, which loves nonrepresentative extraordinary events and shuns normality. When you live on Level 4, everyone on Levels 3, 2, and 1 can look equally poor, and the word *poor* can lose any specific meaning."

Some people have a romantic notion about non-industrialised societies, that they do not suffer the harmful consequences of the modern world or cause damage to their environment. My first-hand experience of extreme poverty was when I visited The Gambia to carry out a feasibility study for a new hospital. This small country is on either side of the River Gambia, and if one digs down a few metres one can obtain fairly fresh ground water for irrigating the subsistence crops that small communities grow. The people seem to live in harmony with their forest surroundings which thrive with bird life, and some people gain employment in eco-tourism. This appears idyllic, and the people (predominantly Moslem) seem contented, but when someone falls ill there is only one hospital in the country, and this has very basic facilities. I asked the hotel manager where I was staying what happens if a tourist suffers a serious heart attack or stroke, and his reply was that they die. There is no effective emergency care service.

The private hospital I was researching was intended to serve the minority rich people in The Gambia and surrounding countries – some of the poorest in the world – and this service would help fund charitable hospital wards. The project was the brainchild of a native of the country who now works as a doctor in the UK. Each year he spends his holidays with friends carrying out free operations on women who have been injured in childbirth such that they leak urine (obstetric fistula), and are rejected by their families because of the smell. There is nothing virtuous about poverty. My friend was lucky because his family could afford to send him to school and he obtained an opportunity from a UK-funded research facility in the country. Most people in the country lack basic education.

Discussion: Readers of this book will probably be in Level 4, or at least Level 3. What level would our grandparents be in? How many generations do we have to trace our family tree back to find peasant farmers who would be in Level 1 or 2 (unless our ancestors were in the aristocracy)?

THE FORMULA FOR ENDING EXTREME POVERTY FOILED BY THE CURSE OF NEO-IMPERIALISM

Rosling: "Today, a period of relative world peace has enabled a growing global prosperity. A smaller proportion of people than ever before is stuck in extreme poverty. But there are still 800 million people left… Unlike with climate change, we don't need predictions and scenarios. We know that 800 million are suffering right now. We also know the solutions: peace, schooling, universal basic health care, electricity, clean water, toilets, contraceptives, and microcredits to get market forces started. There's no innovation needed to end poverty. It's all about walking the last mile with what's worked before."

Our ability to raise ourselves out of poverty has been dramatic. In the year 1800, 85% of the world's population was at Level 1. Even in 1966 it was 50%, but it is now less than 10%. It could be virually eradicated in a decade if the correct policies were adopted.

The UN has carried out many marvellous projects to assist poor countries release themselves from poverty. It was doing very well until, in 2013, the World Bank, under its new president Jim Yong Kim, adopted anti-coal funding policies, effectively prioritising its version of environmental sustainability over poverty reduction. Developing countries couldn't obtain loans for projects with coal-fired energy. The UN 2011 *Sustainable Energy for All* initiative targeted the doubling of renewables' contribution to the global energy mix by 2030, but the cost of this ($500 billion per year rather than $50 billion for coal) contradicts its previous target of erasing extreme poverty by the same year.

In response, Nigeria's finance minister, Mrs Kemi Adeosun, said in October 2016: "We in Nigeria have coal, but we have a power problem, and we've been blocked because it is not green. There is some hypocrisy because we have the entire western industrialisation built on coal energy; that is the competitive advantage that they have been using. Now Africa wants to use coal and suddenly they are saying 'Oh! You have to use solar and the wind' which are the most expensive." There are 600 million people in Africa without electricity, and there will be more than 2 billion people on the continent by 2040.

It seems like malevolent Western imperialism all over again, coordinated this time by the United Nations. When the West prevented the use of DDT in Africa from the 1970s, it effectively imposed a death sentence from malaria on as many people as the murderous regimes of Stalin, Hitler, Mao and Pol Pot. I was told at school that the road to hell is paved with good intentions. We must not make the same mistake again by preventing poor countries using cheap fossil fuels to escape poverty.

In the last decade many of the poorest countries in West Africa like Senegal, Ghana and Mauritania have been exploiting offshore natural gas, to join other countries that are using indigenous oil and gas resources like Nigeria, Angola and Uganda. At the Africa Oil Week in Cape Town in November 2019, Neal Anderson of Wood Mackenzie said: "Most people without access to power are in Africa and the forecasts for population growth are mostly here in Africa. Access to energy is a fundamental human right and Africa is going to be key to growth in energy consumption." In this mood, African countries are not open to suggestions that they stop using fossil fuels. Gwede Mantashe, South Africa's energy minister said: "Energy is the catalyst for growth. They even want to tell us to switch off the coal-generated power stations. Until you tell them, 'you know we can do that, but we'll breathe fresh air in the darkness'."

Environmental groups claim the highest of principles in saving the planet and humanity, but their naïve (green) policies do the opposite. In his book *Golden Rice*, Ed Regis said: "Golden rice has not been made available to those for whom it was intended in the twenty years since it was created. Had it been allowed to grow in these nations, millions of lives would not have been lost to malnutrition, and millions of children would not have gone blind." Peter Beyer (Freiburg University in Germany) and Ingo Potrykus (Institute of Plant Sciences in Switzerland) injected genes of a chemical known as beta-carotene (with an orange-coloured pigment) into the DNA of normal rice, genetically modifying it to provide more vitamin A, which counters blindness and other diseases in children in the developing world.

Patrick Moore: "Greenpeace has opposed the adoption of golden rice… [which] has the potential to prevent the death of two million of the world's poorest children every year. But that doesn't matter to the Greenpeace crowd. GMOs are bad so golden rice must be bad. Apparently millions of children dying isn't. This kind of rigid, backward thinking is usually attributed to the unenlightened and anti-scientific but I've discovered from the inside out that it can infect any organisation, even those as noble-sounding as Greenpeace." Moore was a founder of Greenpeace but left when it adopted irrational policies.

The organisation's website says: "These genetically modified organisms can spread through nature and interbreed with natural organisms, thereby contaminating non-GM environments and future generations in an unforeseeable and uncontrollable way." Claims that GM crops can cause cancer, kidney disease or autism, or harm birds and insects, were dismissed by the National Academies of Sciences, Engineering and Medicine in a 380-page report published in May 2016. It reviewed almost 900 studies on the use and effects of GM crops since the technology emerged 20 years ago. About 12% of the world's arable farming land was planted with GM crops in 2015. More than 80% of the land in soya bean production uses GM varieties and these are widely eaten by livestock in places like the UK. Papayas, bananas and cotton are beginning to benefit from GM technology. Trials are being conducted in China for rice which could increase yields by 50%.

GM crops mean more productivity, less artificial fertilisers, less insecticides and less herbicides. A study by Klumper and Qain at Göttingen University in November 2014 concluded: "On average, GM technology adoption has reduced chemical pesticide use by 37%, increased crop yields by 22%, and increased farmers' profits by 68%… Yield and profit gains are higher in developing countries than in developed countries."

They could be vitally important to poorer countries, but many environmentalists in rich countries oppose them.

Western countries are reluctant to provide chemicals like the pesticide fenitrothion which controls locusts, such that about 25 million people in Tanzania, Uganda and Kenya were suffering food shortages at the beginning of 2020, according to Oxfam, in the largest infestation to be experienced in 25 years. Of the estimated $70 million needed from the UN Food Agriculture Organisation, only $15 million had been allocated by February 2020. In the Americas, the Fall Army Worm larval moth that feeds on maize is kept in check by pesticides and GM strains, but it is claimed that pressure from NGO activists is preventing such techniques being used in Kenya as part of their 'agro-ecology' philosophy. Small scale organic farming is seen as more 'natural' even if traditional 'organic' chemicals like copper sulphate and neem oil are more toxic than carefully researched and tested modern chemicals.

Discussion: Many people today protest about statues of 19th century colonialists whom they say exploited poorer countries to obtain their wealth. If we could ask these original imperialists what their motives were, they might say that they took civilisation to primitive areas of the world, and endowed locals with structured government, modern education, trading systems, railways and other infrastructure. In return they gained profits from their investments. Why do people from more technically advanced countries think they know best what is good for less advantaged ones?

THE BALANCE BETWEEN HUMAN DEVELOPMENT AND ENVIRONMENTAL PROTECTION

European countries did not give much thought to the environment during the Industrial Revolution as the opportunity arose to emerge from extreme poverty. Indigenous forests had already started to decline in the late Middle Ages to build ocean-going boats and battleships. The Agricultural Revolution of the 18th and 19th centuries provided higher levels of food production which allowed more people to go to the cities and work in the new industries. The rural landscape changed, first in Britain, then in other European countries, and then in the rest of the world as human development advanced. Some countries are still destroying wild areas for human habitation and agriculture.

Brazil contains most of the Amazon rainforest, which is still reducing in size as demand for farming land continues. Every so often, tweets appear from world ecology experts like Christiano Ronaldo, Madonna, the Pope, Leonardo DiCaprio ("the lungs of the Earth are in flames") and Emmanuel Macron ("The Amazon rainforest – the lungs which produce 20% of the planet's oxygen"). If they were better informed they would know that the forest plants produce 9% of the world's photosynthetic output but have a zero net contribution to the 20.9% of the atmosphere that is oxygen, which has accumulated over millions of years. The Amazon is a very important ecosystem for a number of reasons, not least that land use change and deforestation is a significant factor in climate change, affecting rainfall patterns, surface albedo (light reflectiveness) and other subtle factors that contribute to regional climate patterns, but it is not "the lungs of the Earth".

In September 2019 Brazilian president Jair Bolsonan, while recognising that more could be done to stop illegal logging, hit back at the UN: "The sensationalist attacks have aroused our patriotic sentiments. Certain countries, instead of helping, embraced the media lies in a disrespectful manner and with a colonial spirit. It is a fallacy to say that the Amazon is the heritage of humankind, and a misconception confirmed by scientists to say that our Amazon forests are the lungs of the world." Some fires were "the ordinary practice of indigenous peoples" on land which had been agricultural for some time (but such primitive methods would stop if the farmers could afford modern farming equipment). France and Germany use "more than 50% of their land for agriculture" but Brazil uses no more than 8%. Closer to home for Europeans, in Romania large areas of forest are being removed by illegal logging that the authorities seem powerless to stop (BBC *From Our Own Correspondent* February 2020). EU bureaucrats may be sitting on furniture made from this timber, while bears, wolves and lynx are losing their habitat. Let's make sure we have our own house in order before we criticise others.

In poorer countries today, the conflict between people and wildlife is complicated and difficult to reconcile. Most palm oil production comes from Indonesia and Malaysia where deforestation is taking place. The people in these countries deserve to earn a living and some deforestation will inevitably take place, but lessons can be learned from other countries to allow some intensive agriculture while protecting a viable area for natural habitats and species. People in rich countries want palm oil in chocolate biscuits, soap, shampoo, cosmetics and cleaning products, but they must be prepared to contribute to the ecological consequences of their consumerism, just as much as the locals. Richard George of Greenpeace UK (*Sunday Times* 3 November 2019) said of supermarket policies: "All big brands say they use only certified sustainable palm oil, or are in the process of switching to it", but he also noted that half of the palm oil coming into Europe goes into biofuel. This is another case of a remedy for reducing emissions causing more harm than good. In the meantime there are only 70 Sumatran rhinos left in the wild and, while there were 175,000 orang-utans in Borneo in the 1950s, they are now critically endangered.

RICH PEOPLE ARE KINDER ON THE ENVIRONMENT

Jesse Ausubel of Rockerfeller University reckons that US corn production has quintupled on the same area of land in the last half century because of modern farming techniques, and that if these were applied globally, an area the size of India could be returned to nature. As more countries got richer and adopted more efficient land use practices, the land area of the world set aside for wildlife in the form of reserves and national parks has grown from 0.03% in 1900 to 0.2% in 1930, and to 15% today. A growing human population has been inevitable, one might say 'natural', but with wealth comes thrift, and an ability to afford measures which protect and nurture the rest of nature. It is in poor countries where people need to hunt ivory to earn a living.

Jordan Peterson: "The data that I've read says that if you can get the GDP of people up to about $5,000 a year, then they start caring for the environment, and the environment cleans up. So you can make a perfectly strong case, I think, and a reasonable one, perhaps even a humane one, that the actual idea would be to get everybody in the world who's poor, desperately so, out of poverty as far as possible." He then criticised the UN for pursuing too many agendas and failing accordingly: "There's 200 millennial goals. That's way too many if you're serious about goals by the way, because 200 goals isn't a plan. It's a wish list. You have to prioritise, but they won't because each of the goals has its constituents, and if you prioritise you irritate the constituents; but if you don't prioritise you can't implement the plan."

Peterson is critical of the UN's policies on climate change because he reckons that reliance on intermittent renewables is not an effective solution, and that we will have to wait for appropriate alternative technologies to develop. In the meantime it would be better to concentrate on other human problems in the world that we know we can solve, which he cites from studies by Bjorn Lomberg mostly relate to poverty and its health consequences. Lomborg calculated that by investing now to obtain the long-term benefits from reduced global warming towards the end of the century, the discounted benefit/cost ratio is a pathetic two cents back on the dollar. He concluded: "Dealing with climate change is a very inefficient way of helping the world." Writing in 2020, Ridley agreed with the poor value for money of current climate change policies: "The IPCC produced two reports last year. One said that the cost of climate change is likely to be less than 2% of GDP by the end of this century. The other said that the cost of decarbonising the world economy with renewable energy is likely to be 4% of GDP. Why do something that you know will do more harm than good?" It would be better to spend the money on human economic development.

A study published in *Nature* on 8 August 2018 by *Song et*

al. at the University of Maryland analysed 35 years of satellite data up to 2016, and found that world tree cover had increased by 2.24 million kilometres, +7.1% relative to the 1982 level. The "net loss in the tropics being outweighed by a net gain in the extra-tropics... Land use change exhibits regional dominance including tropical deforestation and agricultural expansion, temperate reforestation or afforestation, cropland intensification and urbanisation. Consistently across all climate domains, montane systems have gained tree cover and many arid and semi-arid ecosystems have lost vegetation cover."

In reviewing this, Ridley noted that reforestation was associated with a country's stage in economic development. As people moved from the countryside to towns, they stopped cutting down trees for fuel and adopted more efficient and less polluting fossil fuels. Agriculture became industrialised and used less than half of the land to feed the same number of people. Space then became available for reforestation, much of it a sustainable resource to provide indigenous timber for wooden products. "In 2005 twice as much grain was produced from the same acreage as in 1968... Intensification has saved 44% of this planet for wilderness. Intensification is the best thing that has ever happened – from the environmental perspective. There are now over 2 billion acres of 'secondary' tropical rainforest, regrowing after farmers left for the cities, and it is already almost as rich in biodiversity as primary forest. That is because of intensive farming and urbanisation."

In Europe, forest recovery started in the early decades of the 20th century, and in the USA in the 1930s. Brazil is still reducing its forest cover, but countries like Chile and Uruguay have turned the corner. Costa Rica has doubled its tree cover in the last 40 years. Even a highly populated country like Bangladesh is now planting more trees than it is cutting down. Brazil and other tropical countries need to be helped, not chastised, to balance the needs of humans and nature, because they will inevitably reach parity at some point and have the wealth to concentrate more on conservation. Five thousand dollars per person income per year is about $14 per day, which is at the lower end of Level

3, so helping Brazil's economy to improve the lot of those of its population still in Level 2 would be a good thing for the environment. And yet many environmentalists wish to see economic growth halted.

> **Discussion: In forming nature reserves, is it better to exclude human activities altogether and leave nature to get on with it (a truly wilderness area), or have limited human settlement and carefully controlled tourism while managing the environment to maximise its appeal to wildlife?**

THE SLIDE CURVE OF POPULATION GROWTH

Up until the Industrial Revolution, around the beginning of the 19th century, the world's population was fairly stable at about 1 billion people. Each woman would give birth to six children on average, but only two of them made it to adulthood to become parents themselves. Rosling: "There was a balance. It wasn't because humans *lived* in balance with nature. Humans *died* in balance with nature. It was utterly brutal and tragic. Today, humanity is once again reaching a balance. The number of parents is no longer increasing. But this balance is dramatically different from the old balance. The new balance is nice: the typical parents have two children, and neither of them dies. For the first time in human history, we *live* in balance."

If we leave people at Level 1, or even Level 2, we condemn them to an "utterly brutal and tragic" existence. Those who wish to stop economic growth "to save the planet" are swimming against an irresistible tide of human progress. People like David Attenborough and his Population Matters cronies cannot realise how cruel their creed is. He said: "The growth in human numbers is frightening… I've never seen a problem that wouldn't be easier to solve with fewer people." And yet previous futile attempts to stop the 'natural' development of human populations have been inhumane, like India's policy of sterilising 8 million men each year (they were given a transistor radio each in compensation), or China's one-child policy, where district nurses kept buckets of water on hand in case an aborted, 8-month term (illegal) baby was still alive.

As recently as 1965, each woman on average throughout the world bore five children, but the birth rate dropped rapidly as reliable birth control measures became available, and more children survived as treatments for infectious diseases were rolled out (Figure 17). In 2017 each woman had less than 2.5 children. Rosling: "The dramatic drop in babies per woman is expected to continue, as long as more people escape extreme poverty, more women get educated, and access to contraception and sexual education keeps increasing. Nothing drastic is needed. Just more of what we are already doing." There are currently about 2 billion children (up to 15 years of age) in the world, and this has reached its peak. The population will continue to increase until about the end of this century (up to about 11 billion), due to people living longer, with the number of under-30s staying at 4 billion. Europe's population is already starting to drop and the Americas are close to stasis. According to Rosling, the growth will be in Asia (1 billion) and Africa (3 billion), although the recent rapid decline in the birth rate in places like China is driving Asian predictions downwards. The lack of economic progress in many African countries leads to a frustratingly slow decrease in birth rates.

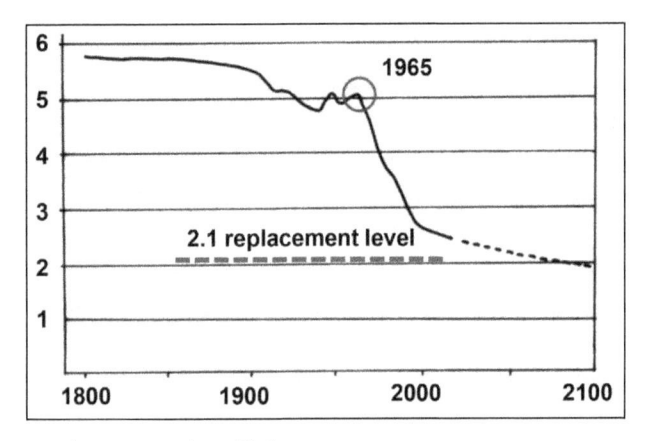

Figure 17 Average number of babies per woman.

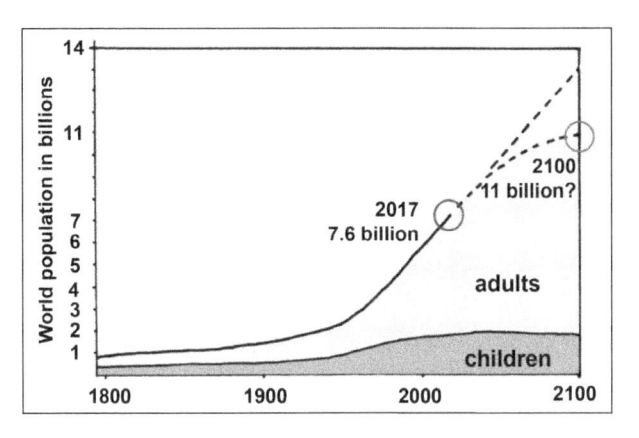

Figure 18 World population. Source: UN and HYDE.

This trend is almost entirely irrespective of religious influence because all advanced Roman Catholic countries now have birth rates of less than two. Italy has only 1.3 children born to each woman (World Bank 2018) and is considering ways to encourage people to have larger families. Even a country like Muslim Bangladesh has, since independence 40 years ago, reduced its birth rate from seven to two with increased wealth. Iran, the country with the world's largest condom factory, has a figure of 2.1.

The graph in Figure 18, up until 2017, is in the shape of what is called a 'hockey-stick', with the 'handle' of the stick up to the mid-20th century and the 'foot' of it thereafter. Hockey-sticks can be alarming because they seem to indicate a feature that is accelerating out of control (the top dotted projection line), but UN projections show a tailing off of the 'foot' due to rapidly reducing birth rates. The graph becomes the shape of a children's slide, called a 'slide-curve'. There is no need to panic. We should accept the inevitable about population growth and take solace in the knowledge that the solutions developed in Level 3 and 4 countries for greater urbanisation, food production and commodities supply can be applied to the increasing populations in Level 1 and 2 countries in Asia and Africa.

THE SLIDE-CURVE OF WEALTH, ENERGY CONSUMPTION AND EMISSIONS

Figure 19 from the US Energy Information Administration (EIA) shows how American energy use rose sharpest in the post-Second World War period as the economy grew, just as it did in the late 20th / early 21st centuries in China. As an advanced economy obtains the standard of living that the West now has, the rate of growth in energy consumption drops almost to zero, the overall graph being a slide-curve. With this comes a levelling off of CO_2 emissions, and then a reduction as gas supplants coal as the chief electricity-generating fuel. In the first decades of the 21st century, America was the country with the greatest emissions reduction due to this.

Note how biomass was a bigger proportion of the energy mix in 1908 due to more trees being cut down, just as bad then for the environment as most biofuels are now. Note also how the use of coal has declined, despite there being plenty of it in the ground, because oil and gas are more practical, affordable and less polluting. Nuclear came into the picture in the second half of the 20th century but is less than 10% of the energy mix because of the high cost of constructing plants.

We get more efficient with what we use. Matt Ridley noted

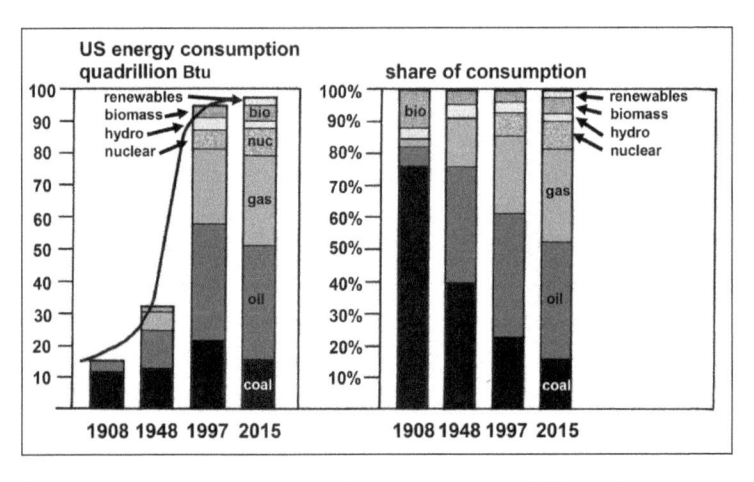

Figure 19 Energy consumption and energy mix in the USA, 1908–2015.

that in developed countries we have reached peak consumption on 66 out of 72 key commodities and their use is declining despite continued economic growth – we are using commodities more efficiently.

All large countries have used coal to develop into an advanced economy. It has high energy concentration, is great value for money and is easily transportable, being in solid form. People tend to suffer the pollution it causes until they become rich enough to afford to use more gas and oil, or nuclear. With this comes a reduction in CO_2 emissions. In 1960 global coal consumption peaked at 20 GJ (gigajoules) per person, and fell back to 15 GJ by the year 2000. Then it suddenly rose to 23 GJ with the development of India and China. Groups like the International Energy Agency expect it to return to the 15–20 GJ range by about 2040.

But the IPCC has models which assume a figure much greater than this. What was originally called the "business as usual" model (RCP8.5 – Representative Concentration Pathway 8.5) assumed oil and gas reserves would rapidly decline in the 2020s and we would turn more to coal. Each person on the planet would use about 30 GJ by 2040, 45 GJ by 2060 and 70 GJ by 2100, and instead of the world's population peaking at 11 billion, it would go to 12 billion. Emissions would rocket.

One decade into the modelling exercise this was proved wrong because of hydraulic fracturing technology, such that we can expect plentiful oil and gas for the rest of the 21st century. While all serious economists recognise that reality is closest to the UN RCP2.6 benign, low-end emissions scenario, the alarmists still make predictions based on RCP8.5. The most likely scenario would mean emissions peaking at perhaps 540 ppm, not the doubling of the pre-industrial level of about 320 ppm to 640 ppm. So if we assume a basic 1.1°C temperature rise for 640 ppm, the effect of a rise from 320 ppm to 540 ppm might be about 0.8°C. Even if we tripled this with feedbacks from the climate system, this would make a rise of just over 2°C the maximum that human emissions could possibly cause, but the comparison of modelling to actuality from 1990 suggests that the positive net feedback is much less than this.

So the lessons appear to be:

- Human population, energy use and emissions follow a slide-curve in a process which seems inevitable.
- We could then let this 'natural' process continue, with the enormous benefit to human well-being worldwide, that is, assuming that the predictions of 'climate chaos' do not transpire.
- Climate chaos is not likely at the lower end of the UN scenarios, which is what we have experienced since 1990.

Discussion: Do wealthy countries need to become more wealthy? In a developed economy are we generally happy with our lives, or is there anything more that money can buy that we need? If we do not need more wealth, why do so many people play the lottery, with millions of pounds to be won?

WEALTH GENERATION AND HIGH ENERGY USE ARE INSEPARABLE

In an ancient hunter gatherer society each person required about 4,000 calories per day to survive, in the form of food and fuel. This was a precarious existence because when the weather/climate changed, from season to season and from year to year, communities had to migrate to another location in the hope that it provided a satisfactory food supply. Migration presents dangers for the more vulnerable people in a tribe, the very young and the very old, and in moving around tribes might come into violent conflict with each other. A new food supply might not be found and all members of the tribe would die. Early agricultural societies provided a more reliable food source, although they did need defence against outsiders who wished to exploit surpluses as they were saved for 'a rainy day', or more likely for when the rains did not come. In such a society, each person could now call on around 20,000 calories per day, not only for food and fuel, but also fodder for domesticated animals, and energy to make weapons and tools. The energy came from wood and peat, highly polluting materials when burned, but tolerable in small societies.

Industrial societies required much more energy, and the fossilised carbon in coal provided this in greater concentration than in wood. Coal can be less polluting than wood but in the huge quantities demanded by energy-hungry societies, smog was a big problem. It was suffered because the benefits far outweighed the downsides. Less polluting and even more energy-rich oil and gas provided the benefits gained in the 20th century, especially those which gave the Western world its comfortable and safe lifestyles in the second half of the century, when nuclear power also contributed to the mix. Coal has roughly one hydrogen atom per carbon atom (with lots of other impurities), refined oil has two hydrogen atoms per carbon one, and methane four (CH_4), to give a more energy-rich fuel. Modern wealthy countries rely on around 180,000 calories per person per day to provide the food, clothing, healthcare, education, transport, law and order, democratic institutions, recreation and a whole range of services and utilities that make our modern lives healthy, safe and enjoyable. Developing nations aspire to achieve such standards and we should help them do so.

Our societies demand energy at all times of the day and night, and at all times of the year, not just when the wind blows at speed and the sun shines brightly. Chapter 10 will demonstrate that wind turbines and solar panels are hopelessly inept at doing this because, in most areas, the wind only blows strongly enough for a minority of the time, and in all parts of the world the sun cannot shine more than half of the time, that is during the daytime. They are no substitutes for fossil fuels and nuclear energy, and we have come nowhere near to inventing a way of storing energy as efficiently as a lump of coal, a can of oil or a flask of methane.

First of all, in the next two chapters, we need to consider what greenhouse gases are, and the part they play in global warming, or at least the current state of knowledge of how sensitive the climate is to increasing human emissions. Part of the IPCC's contention that CO_2 plays a dominant factor in global warming is that they have found no other explanation for it. But there are many scientists who are carrying out research on observed variations and cycles in climate patterns that may have other causes. I look at these in Chapter 9 because, if the problem from human emissions is not as significant as currently officially defined, and natural forces play a bigger part, the policies for addressing climate change might be different.

My generation is much wealthier than my grandparents' one, and the trend is expected to continue, especially in developing countries. The OECD ran models for world GDP per capita from 2010 to 2100 and included them in a report in 2012. The median of the models indicated that the world's population at the end of the century will be 8 times richer, with the lowest model showing 3 times richer. They could even envisage a scenario where a small annual growth accrued in a compound fashion to make people 20 times richer than we are at present. There will be problems to solve, as there have always been, and some of these may be due to anthropogenic climate change, but our children and their children will be more resourceful to deal with them.

Discussion: If the OECD is correct, our wealth will continue to grow over time, especially in currently under-developed countries, despite short-term economic slow-downs that inevitably occur. Apart from dealing with the consequences of future change (there will be always be some downsides of technological and economic development), what good can our increased wealth do in the world during the rest of the 21st century?

CHAPTER 8
THE CHEMISTRY OF 'GREENHOUSE' GASES (GHGS)
CHEMISTRY LESSON

DO WE NEED METAPHORS WHEN FACTS ARE BETTER?
When we walk into a greenhouse on a sunny day, we are immediately aware of how much warmer it is inside than outside. But this is not the 'greenhouse effect' as it works in the atmosphere. There is no glass ceiling in the sky to keep the heat in, or anything equivalent to it. We are also told to imagine a carbon dioxide 'blanket' as if greenhouse gases act as a duvet. They do not. Some scientists seem to think that ordinary people can't understand scientific principles, and that concepts must be simplified for us. They are wrong, and this attitude is insulting to those of us who study the science.

Roz Pidcock, as Head of Communications at the IPCC, wrote the Foreword to *Principles for Effective Communication and Public Engagement on Climate Change: Handbook for IPCC Authors*. She wrote: "Climate change is filled with uncertainties, a notorious stumbling block for communicating with non-scientists. For some the topic can seem abstract and intangible."

"The metaphor of greenhouse gases acting as a 'heat trapping blanket' has tested to be an effective metaphor for communicating the basic principle of the greenhouse effect. More greenhouse gas emissions from burning fossil fuels make the blanket thicker, raising the temperature of the planet. The idea of the atmosphere as a 'bathtub' filling up with carbon dioxide and other greenhouse gases has also been shown to increase comprehension and support for stronger policy action on climate change. The idea to get across here is that even if we 'turn off the tap', existing carbon dioxide emissions will stick around (the bath won't suddenly become empty)."

We do not need to be fed metaphors instead of facts. The bathtub analogy is highly misleading because there is a mechanism for its contents to reduce. Emissions are absorbed into the environment by photosynthesis and other means, although how fast this happens is open to debate. As we saw in Chapter 1, the Leaf Index of the world has been increasing during the satellite era when we have been able to measure it, and this is, in part, due to more CO_2 in the atmosphere.

IPCC metaphors about bathtubs and blankets are not only bad science-teaching and very misleading, they are condescending and manipulative. They want us to "support stronger policy action on climate change", so their motives are political and not purely scientific. But who cares about science? "Research consistently shows that people's values and political views have a bigger influence on their attitudes about climate change than their level of scientific knowledge." Perhaps this is because the public is being shielded from actual scientific facts, and all we have left are political slogans.

"Scientists have been using the metaphor of 'loaded dice' to illustrate that whilst it is still difficult to predict when and where extreme weather events will happen, and though we cannot say an extreme weather event is caused by climate change, we do know climate change is loading the weather dice, making some types of extreme weather events more likely. Be careful not to suggest with the 'loaded dice' metaphor that scientists are 'fixing' their findings to show a certain result." But that is the problem with telling stories instead of facts. As we find out that there is fantasy in the story, we become suspicious of the storytellers.

Discussion: Is it sensible to simplify scientific concepts, or use imprecise analogies? Is the message all that matters? Does the end justify the means, towards the outcome of "support for stronger policy action on climate change"?

WHAT ARE 'GREENHOUSE' GASES AND WHAT DO THEY DO?

The most prominent gases in the atmosphere are: nitrogen (N_2 – 78.084% by volume), oxygen (O_2 – 20.946%), argon (Ar – 0.934%), water vapour (H_2O – 0.4%), carbon dioxide (CO_2 – 0.041% in 2019), neon (Ne – 0.002%), helium (He – 0.0005%), methane (CH_4 – 0.000187%) and Krypton (Kr – 0.000114%). There are other gases in even smaller quantities, like nitrous oxide (N_2O), otherwise known as 'laughing gas' when it is used in concentrated form as an anaesthetic. The volume and mass of H_2O vary considerably in different locations, but 0.4% is the average.

We can spot the 'greenhouse' gases – H_2O, CO_2 and CH_4 – because they have two elements in each atom, as opposed to the others which have only one element (N_2, O_2, Ar, Ne, He and Kr). When radiation from the Sun is radiated from the Earth's surface back into the atmosphere and strikes these two-element gas molecules, and they vibrate as they absorb this infrared radiation, energy is then emitted in all directions. Some of this goes downwards, back towards the Earth's surface, which then warms up. Non-greenhouse gases, with only one element, have no net change in the distribution of their electrical charges

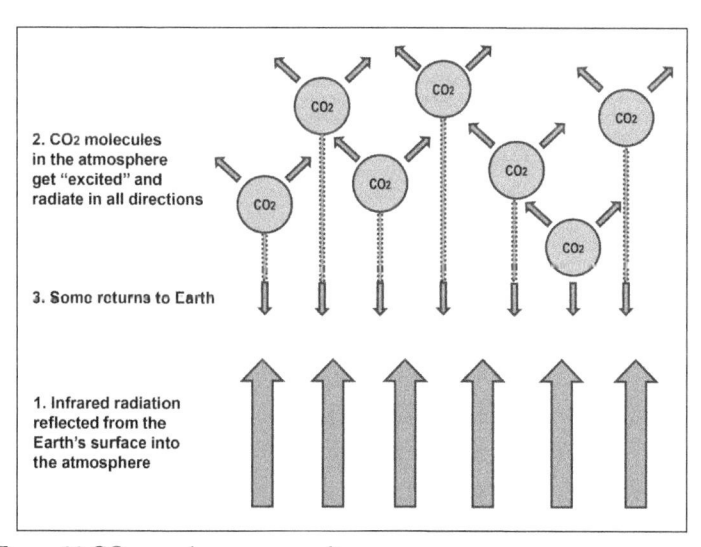

Figure 20 CO_2 greenhouse process diagram.

when they vibrate and are almost totally unaffected by infrared radiation. Ozone (O_3) acts as a greenhouse gas at high altitudes.

Some radiation that returns to Earth from GHGs is absorbed by the surface, in addition to that absorbed directly from the Sun. If it were not for this extra energy, the Earth's surface would be about -18°C. Why is that so difficult to understand such that we have to invent greenhouse roof, bathtub and duvet analogies?

The duvet analogy is wrong because a second layer of 'quilt', or doubling the amount/density of CO_2, does not double the warming effect. It is much less than this. The extra molecules scatter the radiation in various directions in their excitement, including into each other to cause a dampening effect. They don't double the amount that is transmitted back to the Earth's surface.

In Alan Longhurst's 2015 book *Doubt and Certainty in Climate Science* he said: "The effect of increasing concentrations of GHGs, including that of CO_2, is not linear, but logarithmic: a doubling of the CO_2 concentration will not double the radiative effect of that gas in the atmosphere. The radiative effect of the 280 ppm present before the Industrial Revolution in rural areas and at high altitude locations had a radiative effect of about 3.45°C, but to [increase] the concentration to 400 ppm increases its radiative effect to about 3.9°C... The main 14.9 micron absorption band of CO_2 was already saturated in the atmosphere prior to industrialisation." There are also bands where the much more plentiful water vapour is already present such that CO_2 at these frequencies, no matter the quantity, makes no difference.

There are two key points to make:

1. The 'greenhouse power' or effectiveness of GHG's does not increase in proportion to the amount introduced into the atmosphere, but reduces for each 20 parts per million (ppm) added (Figure 21).
2. At certain wavelengths, a saturation point of either CO_2 or H_2O, or both, is reached where no further 'greenhouse gas power' is produced.

I hesitate to introduce an analogy that is often used for this saturation effect, but if one imagines painting a white wall red, the first coat makes the biggest difference until, after a few coats, adding another coat of paint makes no significant difference at all.

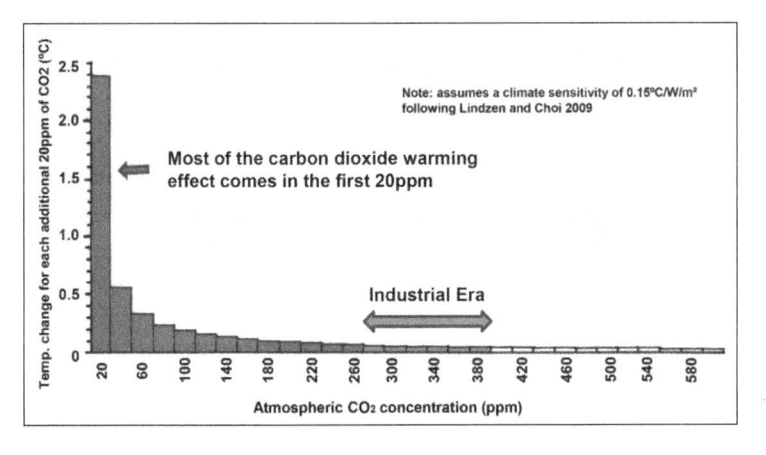

Figure 21 Temperature change for each additional 20 ppm of CO_2.

A good explanation of the saturation of CO_2 at certain wavelengths was given in a lecture on 19 February 2021 by William Happer on YouTube, *How to think about climate change*. Physicist Happer was an advisor to the US Department of Energy in the early 1990s and his career saw him at Columbia and Princeton Universities. He also noted that Greta Thunberg's great-grandfather was Svante Arrhenius (1859–1927) whose intellectual rival in the debate over how more CO_2 emissions might affect temperatures was Knut Ångström. Arrhenius referred to CO_2 as carbonic acid and his formula resulted in a radiative increase due to more of this gas of 5.35W/m². Ångström contested this, noting that Arrhenius did not account for the effect of clouds (more of them due to warming must have a negative feedback because when the sky clouds over it gets cooler) and that the two absorption bands he had identified for CO_2 were already saturated.

Arrhenius was displaying optimism bias because he

recognised that a cool Sweden was not good for the economy, the Little Ice Age conditions having been the default state for half a millennium. He wanted to believe that human emissions would cause global warming and that more CO_2 would fertilise the planet. He wrote in 1908 in *Worlds in the Making*: "We may find a kind of consolation that here, as in every other case, there is good mixed with the evil [coal pollution]. By the influence of the increasing percentage of carbonic acid in the atmosphere, we may hope to enjoy ages with more equable and better climates, especially as regards the colder regions of the earth, ages when the earth will bring forth much more abundant crops than at present, for the benefit of rapidly propagating mankind." More life-enhancing CO_2 certainly has a beneficial effect on Sweden's forestry industry.

WATER IS THE MOST PLENTIFUL GREENHOUSE GAS BUT CARBON DIOXIDE STAYS LONGER IN THE ATMOSPHERE

When we burn natural gas (methane) to produce electricity in a power station, we get the following equation:

CH_4 (methane) + O_2 (oxygen)
> 2 H_2O (two molecules of water) + CO_2 (carbon dioxide)

In other words, twice as much water vapour is produced as carbon dioxide, and water vapour already accounts for about 95% of all greenhouse gases. Why are we more concerned about carbon dioxide than water vapour emissions? H_2O exists in solid (ice crystals and snow), liquid (water) and gaseous (vapour) forms in the atmosphere and is therefore the most dynamic of environmental substances, not only in the lower atmosphere but in the oceans and on land. Some scientists believe that the water cycle is more important for climate change than the carbon (dioxide) cycle, and I will come back to this, but the main reason given for focusing on CO_2 is both its higher radiative property and the fact that it stays around in the atmosphere for a long time. Methane (CH_4) has an even higher radiative power, but there is much less of it.

In other words, while the bathtub analogy is poor, the CO_2 that we put into the atmosphere will build up and take a long time to be absorbed by plants on land and phytoplankton in the oceans by photosynthesis. This is called the 'residency' time of CO_2 in the atmosphere. The IPCC uses the Bern Model to calculate this, and it came to the following conclusion:

- Almost half of newly added CO_2 molecules remain in the atmosphere for only a decade or two.
- Roughly a third of CO_2 emissions stay for a century or more.
- About a fifth of CO_2 emissions stay for a millennium.

Some scientists contest this. The IPCC is again using a theoretical computer model, which climatologists rely on to simulate Earth systems. But wouldn't it be good if we could construct a real, experimental model which was global in size and had all the system elements existing on planet Earth? We could then get results we could trust. But it turns out we have. Until the Test Ban Treaty in 1963 we had routinely set off nuclear explosions in the atmosphere. This caused some CO_2 molecules there to become radioactive, termed Carbon 14 (C14), as opposed to naturally occurring and human produced CO_2 which are called C12 and C13. We could identify CO_2 from different sources and detect how long C14 took to be absorbed into the biosphere. The comparison between this method of calculating residency time and the Bern Model is shown in Figure 22, the former with a ten-year half-life (half gone in this time) and 29 years in Bern. But the latter never seems to go below about 40%.

In 2008 Freeman Dyson debated the subject with Robert May, who had been Chief Scientific Advisor to the UK Government from 1995 to 2000. Dyson had estimated that the total atmospheric carbon is 800 Gt (gigatons) and that photosynthesis absorbs 70 Gt of carbon per year, giving a residence time of about twelve years. A very rough assessment concluded that an increase in 'carbon-eating' plants by a quarter

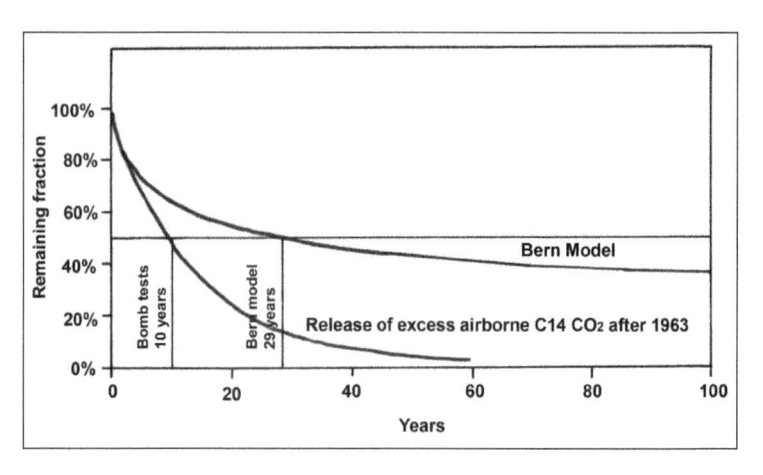

Figure 22 Reduction of atmospheric C14 (nuclear weapon-affected CO_2) over time compared to that assumed for total CO_2 in the Bern Model.

would reduce excess CO_2 in 4 x 12 years, or about half a century. This was very theoretical and does not consider how such immense reforestation could take place, but a lay person can use other logic to conclude that the Bern Model is over pessimistic.

Look at the graph in Figure 23 which shows the increase in atmospheric CO_2 levels since 1959. It is a wavy line because each year the CO_2 level rises by 7.5 ppm and then drops by 6 ppm, giving a fairly constant 1.5 ppm increase each year (presumably due to human emissions). It rises in the summer in the southern hemisphere, which is predominantly ocean and releases CO_2 as the water warms, and drops in the summer in the northern hemisphere, which is predominantly land with more photosynthesising plants sucking it in. It seems that the residency time of the 6 ppm of the absorbed CO_2 is a few months, otherwise the graph would curve upwards rather than being a fairly straight but wiggly trend line.

The Bern Model contends that there is a limit as to how much and how quickly plant life can absorb additional CO_2, and some people contend that while CO_2 levels were much higher, and hence vegetation much lusher, in the days of the dinosaurs than in what some scientists now call our 'carbon dioxide-starved world' (in geological timescales), they claim that plants

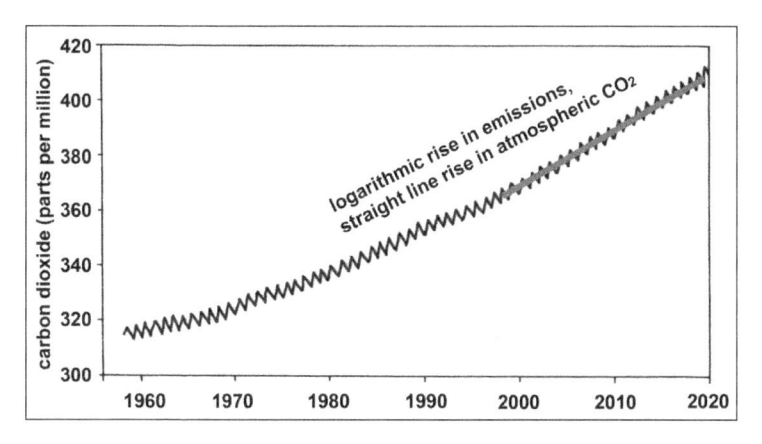

Figure 23 Increase in atmospheric CO_2 from 1959 to 2014 (The Keeling Curve) – but it is not an upwards 'curve' as originally expected in the 1960s, but a fairly straight line.

today have not evolved to absorb unprecedented (in the last two million years) high levels of this gas. However, some farmers pump copious amounts of CO_2 into greenhouses to stimulate growth, suggesting that their appetite for this gas is insatiable. There is also little research into how much phytoplankton growth is increased by more ocean-absorbed CO_2 in waters which rise in temperature to the 5°C to 8°C range that is best for retaining dissolved CO_2.

Discussion: As more countries become industrialised, for example China at the end of the 20th century and the beginning of the 21st, more and more emissions are pushed into the atmosphere. And yet the graph of atmospheric CO_2 does not follow this increasing trend in an upwards curve, but stays as a fairly constant straight line. Why could this be?

HOW MUCH WILL THE AVERAGE WORLD TEMPERATURE RISE IF THE CO_2 LEVEL DOUBLES?

Climate Econometrics: An Overview 2020 by Jennifer L Castle and David F Hendry at the University of Oxford explains the official understanding of climate change in Chapter 4. It quickly jumps from the early demonstration by Eunice Foote in 1856 that greenhouse gases in glass jars warmed in sunlight more than those containing dry air, to make the statement: "Her simple experiment could be demonstrated to school children to explain why CO_2 emissions are causing climate change, leading to a worrying trend in global temperatures." Just because increased greenhouse gases must logically cause some global warming does not justify the statement that the trend is "worrying". There then follows a number of the typical alarmist clichés which might also scare children.

Castle and Hendry seem impatient to skip over the basics to justify their area of research which is, in effect, a combination of economic and climate computer modelling. We do not trust medium- to long-term economic forecasts, and I have argued that climate modelling is a dark science, so combining the two seems to be speculation squared. They often use the word "stochastic" as if this is a scientific term, but it comes from the Greek and means 'guess'. They talk about "empirical climate modelling" which is surely an oxymoron (a figure of speech combining contradictory words). There is very little empirical science in climate change research, but plenty of speculation. It is perhaps unfair to single out this study because its flaws are commonplace in climate change academic research, which is often sophistical (clever but fallacious argument) and unsophisticated (not worldly wise).

I submitted a 17-page review of their paper to Castle and Hendry, what I called an 'un-peer review', pointing out various aspects which I believed were flawed. At least I received a polite reply thanking me for my submission but, as usual, no counter arguments. In my seven years of research into the 'science' of climate change, not a single academic replied to my submissions, even when they invited them after talks they had given. Joseph Romm's 2018 book *Climate Change: What Everyone Needs to Know*, published by the Oxford University Press, is true to its word, only peddling the usual propaganda that it reckons we 'need to know', rather than giving a balanced discussion of what we 'wish to know'. For example, it advocated wind power without mentioning its fundamental flaw of intermittency. I thought of writing to Oxford University, one of our most respected institutions, to point out how the promotion of such biased dogma could severely damage the university's reputation once the truth comes out, but I knew I would be wasting my time.

Universities face an enormous challenge in the Information Age. Before the world of Wikipedia began at the beginning of this century, knowledge gathering was often the preserve of stuffy old professors in ancient institutions. Such is the democratisation of information and its abundance, what is now required of 'official' knowledge gatekeepers is reliable arbitration of data to show its validity, based on what works in practice, and one wonders if universities are 'real-world' enough to fulfil this function. As an employer, I spent a large amount of money retraining post-graduate architects to face the challenges of how to get things done, rather than idly hypothesising about elegant theories impressed on them by academics. My best designer had followed the vocational route to learning.

But I digress. The fundamental question is: how much would the temperature rise if CO_2 went from about 320 ppm in 1960 to 640 ppm at some point in the future? The benchmark was set by the Charney Range, the result of a study by Jule Charney and colleagues in 1979 called *Carbon Dioxide and Climate: a Scientific Assessment*. This estimated a figure for a "transient climate response" of 3°C ±1.5°C; in other words, somewhere between 1.5°C and 4.5°C. That is a wide range, and while activists were happy to quote the higher figure, most scientists reckoned it was most likely to be in the middle to lower half of the estimate.

There are many factors involved. While basic physics suggests that doubling CO_2 should cause a rise of about 1.1°C by itself, the overall climate system has knock-on effects or 'feedbacks' when various factors combine and react with each other. For example, will this 1.1°C rise in warming cause more water to evaporate and the extra H_2O greenhouse effect perhaps triple this figure? Or perhaps more water means more clouds, some of which might reflect sunlight back into space or scatter the infrared radiation, and some of which might retain heat in the lower atmosphere. How can we know? Oceans also absorb and distribute heat as we can see from phenomena like the El Niño, which can change the average temperature over the whole world by more than 1°C in about five years. And how fast will atmospheric CO_2 rise in response to increased human emissions? At present this is about 1.5 ppm/year, which has been fairly constant despite upwardly rising emissions (and an alleged long residency period). At this rate it would take 150 years to reach 640 ppm, even if we ignore the fact that emissions naturally level off and then decline in mature economies. That is, will we ever get to 640 ppm?

From 2012 there was enough data to advance the Energy Balance Method using historical trends rather than purely theoretical computer models. Studies by John R Christy and Richard T McNider in 2017, and Nicholas Lewis and Judith Curry in 2018, gave a best estimate of 1.5°C. This is consistent with a warming rate at the lower end of the IPCC estimate of between 0.1°C and 0.3°C per decade, and Christy's analysis of satellite records currently show 0.13°C per decade (but is some of this natural warming?). At this rate it would be almost 2060 before we reached 1.5°C above the pre-industrial temperature, and 2°C around 2100.

William van Wijngaarden at York University, Toronto, in a lecture on 25 September 2020 to the Irish Climate Science Forum (YouTube *Methane and Climate Change*) estimated that CO_2 will double in about 180 years at an average warming rate of 0.1°C per decade (2.3°C total by the year 2200 with water vapour feedbacks) and CH_4 will double in about 240 years with a warming rate of 0.01°C per decade. At this rate he said that giving up meat would be merely symbolic. He emphasised the uncertainty involved, due particularly to our poor understanding of what clouds do, an area of research he is focusing on. "Thirty years ago the IPCC estimated a climate sensitivity [for CO_2] of about 1.5°C to 4.5°C. After tonnes of work, decades of research, they still have that same temperature estimate." He also notes that there is no clear evidence that atmospheric water vapour has increased in the last century due to a warming climate.

Many academics vehemently object to arriving at what may be an overly optimistic conclusion, especially if it gives an excuse to countries to slow down measures to reduce emissions, because they are not exactly rushing to do so just now. It might also shut down areas of academic research, like econometrics.

Discussion: Should we be wary of people who simplify this complicated subject and claim that there is a consensus among all scientists? Should we plan for the worst possible situation, with all the cost and inconvenience this involves, go for a halfway house, or hope for the best and wait until we can be more sure?

The scientific method is not about imagining possible futures, but about finding verifiable historic data which demonstrates beyond reasonable doubt that a hypothesis is true. We can then use this reality to guide our actions going forward. But even then, our hypothesis may still need to be refined as more data is collected, or it may even be proved wrong and require complete revision. Science never rests and is never "settled", so if someone tells you that it is, or that "the debate is over", you can be sure that they are charlatans, not scientists. Question everything they say.

A big flaw of climate science is that it often lacks the ability to employ empirical science (the real observed world) and has to rely on modelling (speculation about how the real world works). To be reliable models require comprehensive and accurate input data and the means to process 'chaotic' interactions between system elements. It is easy to get the desired results by adjusting inputs, whether intentional or not.

A study by Patrick J Michaels and David E Wojick, published in May 2016, found that climate science, which makes up about 4% of scientific research funding, accounts for 55% of all modelling done in science. Moreover, within climate change science almost all the research (97%) refers to modelling in some way. Roy Spencer: "We don't know the energy flows in and out of the climate system to the accuracy needed to know whether it is naturally in energy balance. So what the modellers do is program the models with the assumption that there is a balance, that is, the assumption of no natural climate change. Then they add CO_2 and the model warms. Then they say, 'see, we've proved that CO_2 causes warming'." Unfortunately, they are not just fooling themselves.

CHAPTER 9
IT'S THE SUN, STUPID
METEOROLOGY LESSON

The climate change establishment has a very high opinion of its research findings because it claims to have uncovered all the possible culprits for global warming. The science is settled, but of course it cannot be by its very nature. Even if critics don't have a clear alternative explanation, it does not mean that there is not one. There are possibilities which we will discuss in this section which are just as credible as the carbon dioxide warming theory, perhaps more so. Those who are formulating them are reluctant to make any definitive claims because they accept that climate science is at an early stage. They are happy to confess their ignorance. They also believe that the conclusions reached by the IPCC were premature because of the political pressure placed upon it in the 1990s. It now has to defend its position and seek ways of reinforcing it. Therefore it is difficult to contemplate new studies which undermine it.

As part of his talk on the subject of deception, magician Matt Dillahuntly recounts the friendship that escapologist Harry Houdini had with Arthur Conan Doyle, who had a fascination with mystery and the paranormal. The former never revealed the secrets of his magic, and Conan Doyle tried to use the logical process of his literary creation Sherlock Holmes to find them. In this, various solutions to a conundrum would be discounted, one at a time, until only one possible explanation remained, however unbelievable. In applying this, he concluded that Houdini must have dematerialised to escape his chains.

As Dillahuntly observed, Conan Doyle could not have known that he had considered all the plausible explanations before he assessed them, and magicians have the ability to disguise the way they do a trick to make the casual observer think that an obvious solution has not been used, when it has. It could be staring us in the face, but our preconceptions might stop our brain from giving it credibility. The same applies to the global warming 'trick'. There may be an explanation for it that our current understanding of science has not revealed to us, or there may be a better explanation of a known factor, like varying solar radiation, that needs further consideration. If we are interested in reasons for a warming Earth, it makes most sense to consider the Sun and its effects on global systems, directly or indirectly.

At the moment, the main flaws in climate science, for which we need answers, are:

- There are various inconsistencies that make the AGW 'fingerprint' unclear; for example, the increase in Antarctic sea ice at the end of the 20th century, the cooling period from about 1940 to 1975 and the pause in the warming from about 1998 to 2014 when emissions were increasing, the troposphere is not warming the way it should if emissions were a dominant factor and periods without human emissions which were as warm as present such as the Roman Warm Period.
- The model-versus-observations discrepancy shows that our current understanding of the sensitivity of the atmosphere to greenhouse gas emissions (the transient climate response) is imperfect.
- The vast volume of water in the oceans affects both the global heat energy balance and the CO_2 budget, but we do not have enough data to understand the processes involved well enough.
- Clouds work to change the energy balance, either to retain heat by the greenhouse effect or to reduce heat by reflecting the Sun's rays, but the mechanisms are not well understood.

THE CARBON CYCLE AND THE WATER CYCLE

Richard Lindzen of MIT has been a critic of the James Hansen view of climate change since he debated him in 1989 at the American Geophysical Union. In 2010 he said: "Our present approach of dealing with climate as completely specified by a single number – globally averaged surface temperature anomaly – that is forced by another single number – atmospheric CO_2 level – clearly limits real understanding. So does the replacement of theory by model simulation. In point of fact there has been progress along the lines [of solar output] and I would suggest that none of it suggests a prominent role for CO_2."

Scientists like Lindzen believe that the answer to climate change lies mostly in the water cycle, not the carbon (dioxide) one, which he believes plays a small part. The driver of the water cycle is not a puny amount of carbon dioxide measured in parts per million, or methane measured in parts per billion, but the enormous fusion power of the Sun and its massive gravitational force. In terms of volume of greenhouse gases, H_2O makes up 95% as opposed to about 3.6% for CO_2. And since human beings contribute only 2.9% of the world's CO_2, this means that humanity's carbon emissions are responsible for only 0.1044% (3.6% x 2.9%) of greenhouse gases. That is a small amount, although those who defend the orthodox position claim it is enough to knock the carbon cycle off balance. Lindzen retorts: "the sensitivity of the climate system to small changes is not an indication of the delicacy of the system, but rather an indication of the ease with which the system can adjust to changes."

When we burn methane to make electricity, twice as much water vapour is produced as CO_2, but satellites fail to confirm that the amount of water vapour in the atmosphere has increased in the last four decades. The ecosystem somehow has a way of dealing with this, or perhaps the amounts released by human gas combustion compared to all the clouds in the sky and all the water in the oceans is so small that we cannot detect the increase caused by us.

We have measured an increase in CO_2 in the atmosphere, going from 0.032% to 0.041% of the air between 1959 and

2019. The orthodox view is that this has been caused by extra human emissions, and some of it definitely must have been. But as we have seen, there is a poor correlation between what we chuck into the atmosphere and the amount actually in it. When Al Gore showed a graph of the planet warming up between ice ages over the last 800,000 years and the CO_2 level going from about 200 ppm to 300 ppm between cold and warm periods, the inference was that CO_2 affected how warm the planet got, even though it was clear that it was changes in the wobble and orbit of the Earth in Milankovitch Cycles that caused Ice Ages about every 100,000 years. The rise in CO_2 followed the warming, not the other way round, the most logical conclusion being that CO_2 was released from oceans as a function of warming.

THE ROLE OF THE OCEANS IN MODERATING ATMOSPHERIC CO_2

Al Gore had a CO_2-centred state of mind which made him focus on it in a cause-and-effect explanation of historic changes in atmospheric CO_2, just as the geocentrist state of mind of religious authorities in the 16th century, where man is made in the image of god and everything revolves around us, could not accept the work of Nicolaus Copernicus (*de revolutionibus orbium coelestium* 1543) when he demonstrated that the Earth revolves around the Sun. With a CO_2-centred state of mind, it is this gas, increased exclusively by (greedy) mankind, which affects the oceans, without considering that perhaps it is a two-way process, and that the oceans have the bigger say.

Could it be that the warming in the 19th and 20th centuries, in a 'bounce-back' from the Little Ice Age, was caused mostly by non-human forces, and this led to increased CO_2 as the warming oceans gave it up? We know that human emissions are increasing in the atmosphere, but the enormous surface area of warming oceans releasing CO_2 into the air could explain why the Keeling "Curve" (Figure 23) is effectively a straight line. Perhaps the CO_2-releasing effects of the oceans are evening out the stepped temperature rises since the end of the Little Ice Age to an almost steady state CO_2 upwards trend. Direct human emissions would be small in comparison which would explain why the straight line rise in atmospheric CO_2 does not correlate well with the upwardly rising curve of human emissions. The dominant, buffering power of the oceans would also explain the virtually straight line rise in sea levels.

We should be wary when two data sets are joined together, as they were with official records of atmospheric CO_2. Ice core sampling of the gas is taken up to 1959 when chemical measurements began at the Mauna Lua Observatory in the middle of the Pacific Ocean in Hawaii. But surely we took chemical measurements before this date? Flemish chemist Jan Baptist van Helmont first identified CO_2 about 1640, and Scottish physician Joseph Black defined its properties in the 1750s. Humphrey Davy and Michael Faraday first liquefied it in 1823. The science of carbon dioxide is a lot older than the science of climate change. It seems that we have been measuring atmospheric CO_2 since long before David Keeling was just a glint in his father's eye.

Beck et al. 2009 (Figure 24) reviewed 170 documented sources and found 90,000 historic CO_2 measurements from 1812 to 1957 and they reckoned that the readings from about 1870 had a low margin of error. These showed a general background level of about 310 ppm during the Industrial Era, with a peak of about 380 ppm during the global warming phase of about 1920

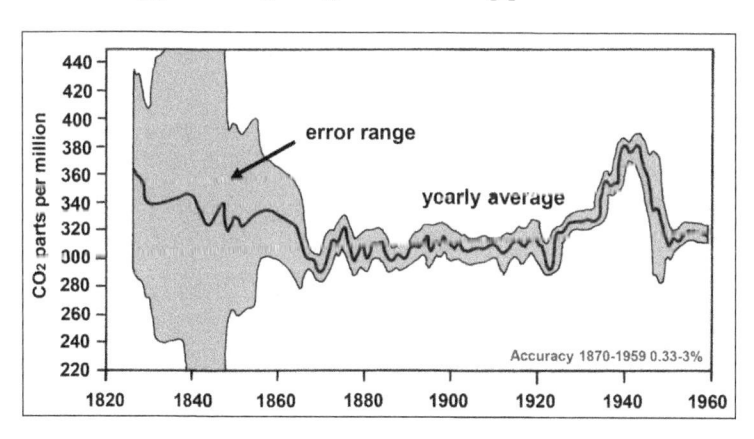

Figure 24 Atmospheric CO_2 from 1826 to 1960. Source: *Beck* 2009.

–1945. *Kouwenberg et al.* 2005 appeared to confirm this when they reconstructed paleo-atmospheric CO_2 from fossilised pine needles for the last 1,800 years and found three peaks up to about 350 ppm.

If we are now heading to about 415 ppm, is it the extra 35 ppm (above the 1940 peak of 380 ppm due to the warming phase from 1920 to 1945) that post-1950 human emissions are directly responsible for, with the rest caused by the natural global warming which took place between 1975 and the end of the century?

Due to Knudsen Diffusion (microscopic gas bubbles escaping through tiny cracks in the ice core when it is released from the high pressure of a glacier), is ice core sampling only sensitive enough to indicate a long-term CO_2 level of about 280 ppm, and not a short-term, decadal one to pick up spikes like the one shown in Figure 24 in the 1930s and 1940s? If the CO_2 level dropped from 380 ppm to 320 ppm between 1940 and 1950, does this show that the residency period of this greenhouse gas is about ten years? Surely this indicates that increased temperatures and CO_2 levels do not cause tipping points? Or is there something flawed in the analyses of Ernst-Georg Beck, Merian-Schule Freiburg, Lenny Kouwenberg and others?

Kouwenberg et al. also quoted *Fischer et al.* 1999 and 2004 work by EPICA members to support their pine needle (stoma) evidence of periodic peaks (350 ppm) and troughs (230 ppm) in CO_2 levels, which have some correlation with the rising and falling temperatures in the Roman, Dark Ages, Medieval and Little Ice Age periods. Kouwenberg saw sea temperatures as the key to understanding the rise and fall of CO_2 levels. "Involvement of sea surface temperature changes in the production and depletion of atmospheric CO_2 is strongly suggested by the apparent synchroneity between the timing of CO_2 maxima and minima in the stomata-based record (within certain limits), and changes in North Atlantic Ocean SST as recorded offshore the Mid-Atlantic United States (*Cronin et al.* 2003)... CO_2 fluctuations over the last millennium at least partly could have originated from temperature-driven changes in CO_2 flux between ocean surface waters and atmosphere." That is, some CO_2 is entering the atmosphere because of warming oceans.

OUR POOR UNDERSTANDING OF THE OCEANS

Despite the efforts of the many IPCC contributors, there are lots of things we still do not understand about clouds and oceans. The Argo programme was started in 1999 to address the latter problem. For four centuries, sailors have understood the two-dimensional ocean winds and currents that set the trade routes on which they travelled on their sail ships. Argo buoys are beginning to improve our understanding of the three dimensional flow of currents that absorb, transfer and release heat around the world, and are players in climate changing phenomena like El Niños. There are now 3,800 floats that measure the temperature, velocity and salinity of the upper 2,000 metres of the ocean. This is still a small number in relation to the vastness of the oceans, but it is a start.

Argo's website states: "Lack of sustained observations of the atmosphere, oceans and land have hindered the development and validation of climate models. An example comes from a recent analysis which concluded that the currents transporting heat northwards in the Atlantic and influencing Western European climate had weakened by 30% in the past decade. This result had to be based on just five research measurements spread over 40 years. Was this change part of a trend that might lead to a major change in the Atlantic circulation, or due to natural variability that will reverse in the future, or is it an artefact of the limited observations?"

In private, climate scientists understand where their knowledge is weak, but in official statements they cannot afford to be candid. This admission is from Kevin Trenberth, of the University Corporation for Atmospheric Research, and a lead author of IPCC Assessment Reports. It was in a (leaked) email of 14 October 2009 to Michael Mann. Following the realisation that warming had stopped since 1998, Trenberth stated: "How come you do not agree with a statement that we are nowhere close to knowing where energy is going or whether clouds are changing to make the planet brighter? We are not close to balancing the energy budget. The fact that we cannot account for what is happening in the climate system makes any consideration of geo-engineering quite hopeless, as we will never be able to tell if it is successful or not… Saying it is natural variability is not an explanation. What are the physical processes? Where did the heat go? … The resultant evaporative cooling means the heat goes into the atmosphere and should be radiated to space: so we should be able to track it with sky temperature data. That data is unfortunately wanting, and so too are the cloud data. The ocean data are also lacking, although some of that may be related to the ocean current changes, and burying heat at depth, where it is not picked up."

THE CLOUDS AND THEIR EFFECTS ON THE WORLD'S ENERGY BALANCE

Fundamental to the construction of computer models is an understanding of the world's energy balance. A diagrammatic representation of this is shown in Figure 25. Three hundred and forty watts per square metre (W/m^2) of solar energy arrives as shortwave, visible radiation and goes through various global interfaces, but the main thing to note is that it is assumed that everything is in perfect balance until human activity is introduced. This causes an imbalance such that the Earth's surface warms by 0.6 W/m^2 (bottom right corner of diagram).

In Figure 25, the figures in the arrows have been rounded to the nearest whole number, so the imbalance is 1 W/m^2, but 0.6 W/m^2 is the figure used. How can the originators of the diagram be so precise? The solar heat reflected off clouds back into space is a neat 100 W/m^2, perhaps too neat to be credible, but suppose the average cloud cover varies by a not unreasonable 2% over a year or a decade, then the resulting 2 W/m^2 change completely overshadows the 0.6 W/m^2 imbalance allegedly caused by humans. But clouds vary with time and air pressure, they do not stand still for even a second, such that their opacity/

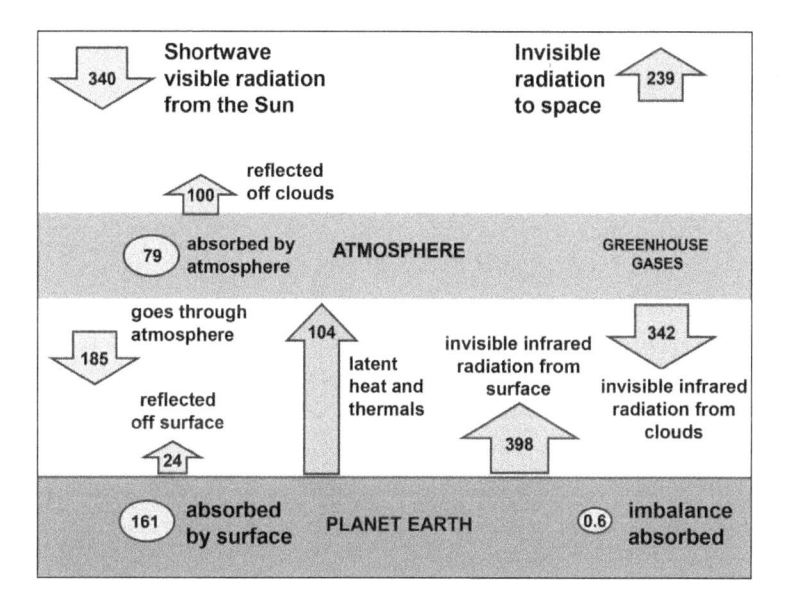

```
┌─────────────────────────────────────────────────────────────┐
│  ↓ 340   Shortwave              Invisible       239 ↑       │
│          visible radiation      radiation                   │
│          from the Sun           to space                    │
│                                                             │
│               reflected                                     │
│        ↑100   off clouds                                    │
│─────────────────────────────────────────────────────────────│
│         ⎛79⎞ absorbed by  ATMOSPHERE        GREENHOUSE      │
│              atmosphere                     GASES           │
│                                                             │
│         goes through                                        │
│         atmosphere    ↑104                       ↓342       │
│  ↓185                      invisible infrared               │
│                       latent radiation from                 │
│         reflected     heat and surface       invisible infrared│
│         off surface   thermals               radiation from │
│                                      ↑398    clouds         │
│         ↑24                                                 │
│─────────────────────────────────────────────────────────────│
│         ⎛161⎞ absorbed   PLANET EARTH   ⎛0.6⎞ imbalance     │
│               by surface                      absorbed      │
└─────────────────────────────────────────────────────────────┘
```

Figure 25 Global energy balance diagram typically used by climatologists. Units in W/m².

reflectivity cannot be measured, only very roughly estimated. The arrogance of this simplistic diagram is startling, suggesting the possibility of an average energy model for the whole planet, with all the constantly changing cloudy and cloudless world zones. There are variables which are impatient of definition. The scenario shown may be correct at one moment on one day of the year, but the next it will be different.

I have a hunch that the key to filling in critical gaps in our understanding of climate change is the effect clouds have on the overall system. When the sun goes behind a cloud on an early summer's day, there is an instant chill in the air, and often a gust of wind. The last time I crossed the Atlantic in a plane, there was cloud cover the whole way from America to the UK, as far as I could see towards the Arctic Circle. The last time I flew from Amsterdam to Cape Town, I could not see the ground from north of the Tropic of Cancer until dusk around the Tropic of Capricorn. We live more on a white planet than a blue one.

THE SUN AS THE SOURCE OF ALL HEAT HOLDS THE KEY TO UNDERSTANDING GLOBAL WARMING

Apart from heat from the Earth's core emitted via volcanos, the Sun provides the only source of warmth to the planet's surface. If it is warming, perhaps the first place we should look to is our home star. The Sun's radiation varies constantly, but only by about 0.1%, so the IPCC and many scientists dismiss solar energy changes as a significant cause of global warming since 1820. But others argue that small changes can be amplified by mechanisms within the Earth system, most particularly the clouds.

We have been measuring sunspots for hundreds of years to observe that there are periodic bursts of solar flares (coronal mass ejections), but we have only recently realised that cosmic rays from beyond the solar system play a part in seeding clouds and, when the Sun's activity and gravitational forces are weaker, more cosmic rays get to Earth and more clouds are seeded. This concept is best explained by Nir Shaviv of the Hebrew University in Jerusalem, although a lot of the research has been carried out by Henrik Svensmark at the Danish National Space Institute, whose first paper on the subject appeared in 1997.

Shaviv recognised that the IPCC predictions for the doubling of CO_2 had been in the range 1.5°C to 4.5°C since before 1990, but that observations were at the bottom edge of this. *Douglas et al.* 2008 showed that the warming trend should be greater higher up in the atmosphere than at the Earth's surface, according to the received wisdom of the time, but that this was not the case. There was also no good 'fingerprint' match for CO_2 and warming. He was suspicious of the Michael Mann hockey-stick of historic temperatures (more of this in Chapter 12), or "hockey stitch" as he called it, because it joined together two different datasets, proxy data from tree rings and measured data from thermometers. He noticed that "different 'recipes' for the cloud cover produce different sensitivities". In other words, various assumptions on when clouds acted as a greenhouse-effect agent and when they acted to block the Sun and reflect heat back into space gave a wide range of answers. *Cess et al.*

1989 concluded that this was the single most unknown factor in climate science.

The answer to the various conundrums must lie in the varying power of the Sun. About every 11 years the Sun flips polarity, and there are periods of longer scale mass ejections which we notice because there are radio communications disruptions on Earth. *Neff et al.* 2001 measured a proxy of solar activity from tree rings to detect Carbon 14 which had been formed by cosmic rays. These hit the atmosphere and can break nitrogen and oxygen nuclei into smaller particles of C14. Cosmic rays are modulated by solar winds which, if stronger, stop more of them entering the atmosphere.

The role of changing cloud cover had been considered as far back as *Ney* 1959, and *Dickenson* 1975 had predicted what *Svensmark* 1998 observed. It seemed that when there was less ionisation, the clouds were less white, there was less reflection by them and the Earth warmed. *Marsh & Svensmark* 2000, *Shaviv* 2002, *Shaviv & Veizer* 2003, and *Svensmark et al.* 2009 advanced the research. Svensmark could show that a gust in solar wind led to a reduction in the cosmic ray flux reaching the Earth several days later, with a reduction in aerosols and cloud properties observed by satellites, and he correlated the five strongest Forbush decreases of cosmic ray flux between 1987 and 2007 with changes in the clouds, covering two 11-year solar cycles. (The Forbush rapid decrease in cosmic rays following a coronal mass ejection was named after the American physicist Scott E Forbush who identified this phenomenon in the 1930s and 1940s.)

Svensmark demonstrated the change in the ionisation nucleation rate of condensation by cosmic rays in an ionisation chamber in Stockholm, and this was repeated at CERN by Jasper Kirkby and colleagues in their CLOUD experiments (Cosmics Leaving Outdoor Droplets), with a proton synchrotron producing artificial cosmic rays. This was reported in *Nature* on 25 May 2016. The study concluded that climate scientists had previously exaggerated cloud formation seeded by atmospheric pollution like sulphuric acid and ammonia, and underestimated the amount of cloud cover in pre-industrial times. Climate models should be recalibrated accordingly, and this would reduce the warming attributed to anthropogenic factors.

Shaviv 2008 found a correlation between solar flux and the sea level change rate during the period 1920 to 2000. While the sea level graph in Chapter 3 (Figure 13) shows a straight line trend, they contain peaks and troughs which no one had been able to convincingly explain before. Shaviv calculated a 1.0W/m^2 change in solar flux which correlated with the difference between peaks and troughs, greater than the 0.6W/m^2 that Figure 25 attributes to human activities. The IPCC modelling had assumed that solar radiance variation could only be about 0.17W/m^2, based on measurements of direct radiation from the Sun, failing to recognise the amplifying effect of changing cloud conditions. Shaviv reckons that such modelling therefore underestimates the effects of solar radiance by a factor of six or seven.

Ziskin & Shaviv 2012 plugged the new radiance figures into IPCC models and achieved a margin of error against 20th century recorded data of ±0.1°C, against the IPCC margin of error of ±0.2°C. They then integrated this into predictive models and calculated that a doubling of CO_2 in the atmosphere would lead to a 1°C rise in temperatures, less than the IPCC minimum prediction. Shaviv is reluctant to shout his message from the rooftops because, while we observe that cloud particles can be seeded by cosmic rays, we still don't understand the mechanism for how this happens. But lack of knowledge is not as bad as claiming knowledge where uncertainty exists. "What gets us into trouble is not what we don't know. It's what we know for sure that just ain't so." As for making predictions of what the climate will be like in the future, Shaviv declared, "I am not a prophet, even though I come from Israel."

The contention that the change to a single element in the climate system, a 40% increase in the trace gas CO_2, has an overriding effect on the whole system is difficult for me to believe. Surely the very powerful water cycle has more effect on climate than the carbon cycle? Indeed, to achieve the 1.5°C+

rise predictions, the carbon cycle (doubling CO_2 by itself only has a theoretical 1.1°C rise) has to borrow power from the water cycle and make it worth bothering about. The orthodox IPCC hypotheses have been pretty well fixed since the early 1990s, despite the fact that climate models based on them have consistently exaggerated the warming that the world has experienced over the last three decades.

STRATOSPHERIC WARMING AND ITS EFFECTS ON THE LOWER ATMOSPHERE

There might be another or complementary explanation for how changing solar radiance is amplified in the atmosphere. Joanna D. Haigh, Professor of Atmospheric Physics at Imperial College London, in a 2017 lecture on YouTube, outlined her studies into the changes in temperature and ozone concentrations in the stratosphere which seem to correlate with the 11-year solar cycles. The variation of radiation at the Earth's surface at the visible end of the spectrum, where most of the energy is transmitted to the planet, can, by itself, only account for one or two tenths of a degree centigrade difference in temperatures. But she observed that, in the upper atmosphere, where radiation is at ultraviolet wavelengths, down to 100 nanometres, the temperature can double. Perhaps there is a knock-on effect on surface temperatures from the upper atmosphere via clouds and oceans.

"Current work is focusing on coupling between the different layers of the atmosphere, the stratosphere and the lower atmosphere, the troposphere, and also the coupling between the troposphere and the oceans as to how some of these apparent solar effects take place. We can now understand that changes in the stratosphere can affect the surface climate through changes in the winds and the circulations, and then there may be a knock-on effect of the wind stress on the oceans, producing some of the observed ocean effects… In summary, changes in solar radiation reaching the Earth, either because of the Sun's variation, or because of the Earth's orbital variation, is fundamental to our understanding of climate and climate change, and we need to understand that in order to be more certain about human activity and how that's affecting the climate... The [satellite] technology is not quite good enough yet but it's developing all the time." In other words, we have lots of theories which need to be proved or discounted, but we do not have enough data because the science is at such an early stage.

Haigh, like Shaviv and Svensmark, has a hunch that the key to understanding how the planet warms lies in the Sun, the only source of our heat, apart from the stored heat from the Earth's core (including from fossil fuel extraction and burning) which is tiny in comparison. She is cautious about jumping to conclusions. For example, she recognises that there is a good correlation between a 50-year depletion of sunspot activity and the Maunder Minimum (the coldest point in the Little Ice Age around the year 1700), but notes that there was a lot of volcanic activity at the time which may have contributed to the cooling by ash particulates in the atmosphere blocking out the Sun's rays. But some scientists speculate that solar and planetary gravitational forces, acting on the Earth's crust, trigger seismic activity, as well as affecting how many cosmic rays penetrate the atmosphere and seed clouds. We have a lot still to learn.

One of the best TV documentaries I have seen which tried to determine the causes of severe weather events was from the BBC Horizon series in 2014, *What's Wrong with Our Weather?* Presenters John Hammond and Helen Czerski set off with an open mind and explored factors which changed the path and speed of the Jet Stream, which determines the weather patterns affecting northern Europe and North America, either dragging in cold or warm air, the former having been the trend in the previous few years. They explored tropospheric events in Asia and the Caribbean which affected the Jet Stream's path, as well as the influence of the stratosphere. They considered the loss of Arctic ice but played down Arctic Amplification as a determining factor because data had only been available for 30 years. The only influence they reasoned from global warming was more moisture in the air causing heavier downpours, a theory they said "had not yet been disproved". They pointed out that we look for patterns in climate but, just like observing heads coming up three times in a row when a coin is tossed, this does not tell us what will happen the next time. The same applies to the weather and climate because so many factors are present. We need a popular, competent study like this every five years to update our knowledge of current science.

DARE THE IPCC EVER ACCEPT SUCH RADICAL NEW SCIENTIFIC EVIDENCE?

IPCC AR6 was published in August 2021, too late to include in the body of this book, although I would probably have started with *IPCC AR5* anyway as a base point and tracked IPCC strategy over the last seven years. It takes several months to read and analyse these lengthy and reader-unfriendly documents, and as Chapter 12 demonstrates with *IPCC AR3*, it can be several years before flaws in a report can be exposed. So my review of it in due course will reveal what appears in it about the cosmic ray hypothesis or the connection between cyclical stratosphere temperatures and surface temperatures, if anything, and what it says about the measured rate of warming against the theoretical one.

It would be extremely embarrassing for the IPCC if it were to admit that the problem is not as severe as predicted, just as the political momentum for urgent action has been wound up, albeit with little actual response from countries for the reasons we have discussed, 'hot air' being more in evidence than 'cool action'. It clearly believes that action is necessary, so it cannot take its foot off the pedal. In this political climate, a clean-slate reassessment, based on all the knowledge we currently have, in a Green Team/Red Team balanced scenario, does not seem possible.

Too many people have invested too much of their careers and reputations to embrace a new mindset. The momentum of the enormous eco-industrial complex, and the feast of researchers on climate change-related academic research budgets, would be almost impossible to stem. Steve Koonin: "Fifteen years ago, when I was in the private sector, I learned to say that the goal of stabilizing human influences on the climate was 'a challenge', while in government it was talked about as 'an opportunity'. Now back in academia, I can more forthrightly call it what it is: 'a practical impossibility'... I understood the science and the societal changes well enough to see that a straightforward synthesis of a handful of basic facts led directly to the conclusion that even stabilizing human influences was so difficult as to be essentially impossible. Saying that directly and publicly while I was working for BP, and later the Obama Department of Energy, would probably have gotten me fired."

What is worse is that Donald Trump would have been proved right, even if his conclusion had been reached by instinct rather than intellect. It was fake news all the time. If there is something that is more distasteful than Al Gore arrogance, it is Donald Trump smugness.

CHAPTER 10
WHAT CAN WE DO TO REDUCE HUMAN EMISSIONS?
TECHNOLOGY AND ECONOMICS LESSONS

So much fuss had been made by climate change activists, who claimed the authority of the United Nations, that politicians felt they had to respond by doing something. They could not be accused of failing to 'save the planet'. The rich countries, which could afford to do so, started to adopt policies aimed at reducing CO_2 emissions, without working out how much they would cost or even how well they would work. This is a long chapter and perhaps the most important because it is here that theory hits reality – how to implement a net-zero emissions policy – which seems to me, with existing technologies, to be an impossible target.

THE INCESSANT MARCH OF FOSSIL FUELS TO POWER HUMAN PROGRESS

Figure 26 shows the world's energy mix from 1970 to today, with projections until 2040. Since energy projects take a long time to develop, we can make fairly good estimates of energy use and supply for a decade or two in advance. The use of coal, oil and gas will continue to grow as the population increases towards the 10- to 11-billion peak and poorer countries strive to achieve the same level of wealth that Europe, North America, Japan, Australia, New Zealand and some other countries have just now. At the beginning of the 21st century the contribution from renewables (wind, solar and biofuels) was barely perceptable in world terms, and by 2040 the modest increase will barely keep

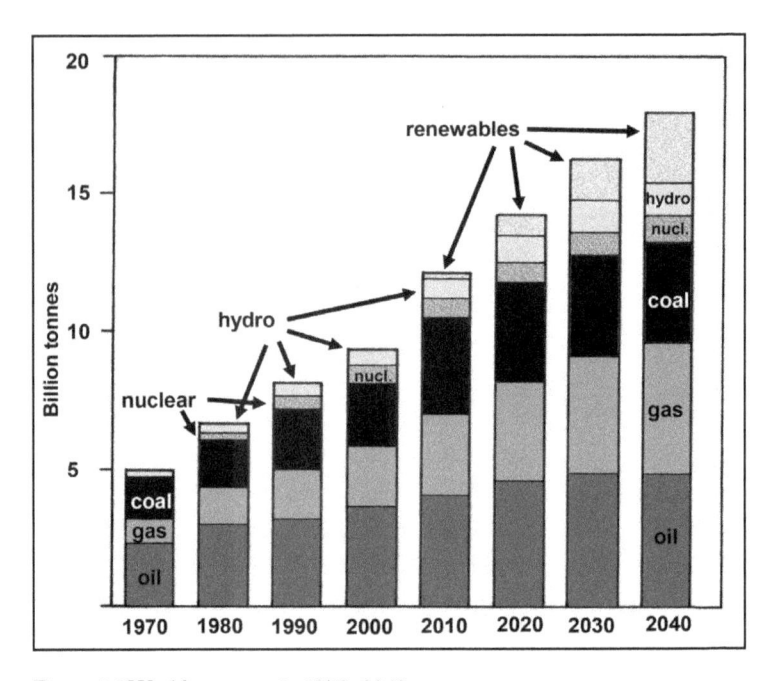

Figure 26 World energy mix 1970–2040.

up with increased energy demand. CO_2 emissions will inevitably increase to 2040 and probably for many decades beyond this. Any savings by rich countries, whose emissions naturally level off and reduce by about 30% anyway, will be swamped by rises in developing nations.

WHAT DO ALTERNATIVE ENERGY SOURCES NEED TO DO TO MATCH UP TO OIL AND GAS?

At the moment, wind power is the biggest tool in the emissions reduction armoury, with solar being used mostly in countries which get a lot of sunshine. There are lots of people trying to develop other alternative energy systems to fossil fuels, but until they can be demonstrated to be capable of being rolled out at a scale large enough to satisfy demand, and are affordable, they cannot be included in any alternative energy plan. An energy source must pass a number of viability tests if it is to supplant oil and gas:

1. How efficient and reliable is it compared to oil and gas?
2. How does the cost compare to oil and gas?
3. How does its use affect the environment?

Before I consider whether wind and solar are up to scratch, I will give a brief explanation of why other technologies and strategies currently being considered cannot provide energy in the quantities required and are often very harmful to the environment they are supposed to be saving.

Hydro is only applicable on a large scale if the country involved has a lot of mountains and lakes, like Sweden, and if it is deemed ecologically acceptable to dam rivers and flood valleys. Mostly, technologies like hydro, tidal and wave power can only supply a very small amount of a country's energy requirements.

The UK's Centre for Alternative Technology think tank discounts hydrogen as a domestic fuel for various reasons discussed later in this chapter, and instead reckons that we need to obtain 75% of domestic heating and hot water from heat pumps. Heat energy naturally transfers from warmer places to cold ones, but heat pumps reverse this process by using a refrigerant to absorb heat from the air or a ground source, to vapourise it, then release the heat as it condenses. The heat pump requires electricity to operate but it releases more heat than it consumes, the coefficient of performance (COP). There are however practical issues. Ground source heat pumps generally require a garden up to three times the floor area of the house to install ground loops. Any trees in the garden would have to be removed so their roots do not harm the loops. There also needs to be sufficient space to install the alternative air source heat pumps externally, away from neighbours' properties and preferably in a southern-facing sunny spot. There must be space in the house for a buffer tank and water cylinder. Underfloor heating works better than radiators, which must be oversized compared to traditional systems. Heat pumps work poorly in very cold weather, taking a long time to heat up a house, in contrast to the flicking of a switch on a gas boiler which makes a home cosy in minutes. Even if part of the installation cost (£15,000+ per home?) is covered by government grants, many homes will not have the space and the disruption during installation will be enormous. What government dare impose this on people?

I have already been critical of biofuels because they use large quantities of fossil fuels to cultivate, crop, process and transport to locations they are used, despite the UN classifying them as carbon neutral, which they clearly are not. Not only are they ineffective in saving emissions, they are extremely harmful to the environment. Large areas of forest in North America are cut down (equivalent to 24 million trees each year), with huge ecological damage, the wood is dried and chopped into pellets and it is transported across the Atlantic to be burned at the Drax power station in Yorkshire to generate electricity. Many European countries use palm oil from Indonesia and Malaysia as biofuels, which again results in huge ecological damage. Even if we believe that the end result is carbon neutral, this only happens after about 40 years when the forests fully regenerate and take CO_2 from the atmosphere in the quantities they did before they were harvested.

Bioenergy crops demand an unrealistically large amount of space. David MacKay (*Sustainable Energy Without the Hot Air* – free online) calculated that to provide enough biofuels to power cars, there would need to be an 8 km-wide strip of vegetation the whole way alongside each major road. If we have spare land to devote to biocrops like maize, it would be more ecologically sound to set this aside for nature reserves or for forestry. Greenpeace no longer supports biofuels, unless the materials used are purely waste products such as offcuts from forestry management, but these do not produce wood in sufficient quantities.

However, many politicians fell into the biofuels trap. The UK taxpayer is bailing out the Northern Ireland Renewable Heat Incentive that was supposed to cost £25 million in its first five years but is likely to reach £1.15 billion over 20 years. Farmers heating empty barns in order to obtain public subsidies is plain stupid. No doubt those who set up the scheme thought they were saving the planet, but they are just making us all poorer. It also caused a political crisis in Northern Ireland, the 'cash for ash' crisis that brought down the government there and reinstated rule from Westminster for three years, which in effect meant no rule since London did not wish to risk becoming embroiled in Belfast political sensitivities it didn't understand.

> **Discussion: Many people have installed wood burners in their homes (wood is a biofuel), believing that they are more 'natural' and comforting than oil or gas-fired central heating, but each one produces the particulate emissions equivalent to 18 diesel cars. Is this OK, particularly in urban areas? How much should it be left to individual citizens to decide what is best for them and their neighbours, and how much should the state interfere?**

NUCLEAR POWER

Nuclear power should be a significant part of many countries' energy mix, but it is expensive at present, especially the capital costs of the huge and complex buildings. If we did not have to raise the capital to begin with we could justify the whole-life cost of nuclear because we can design power stations for a 60-year life or more. And they are not a quick fix. If they take perhaps 12 to 15 years to design and build, and another six or seven years to win back the CO_2 emissions involved in construction, a facility started in 2020 would not start to save emissions until around 2040.

Nuclear only provides a base load because the output from a power station cannot be varied quickly to meet demand as it changes at different times of the day and the year. Gas power stations afford this flexibility, which is largely why they are so popular. Future developments in nuclear energy technology may make this energy source more affordable, but it is unlikely that they will provide more than just a base load; in other words, the minimum amount required most of the time, not the peaks in demand.

The German government was persuaded by the Green Party to close all the country's nuclear power stations after a tsunami damaged the Japanese power plant at Fukushima in 2011, despite the safety features there ensuring that no one died from radiation poisoning (some people claim there was one indirect death). To keep the lights on, German coal generators with large CO_2 emissions were kept working, and pipelines were built to bring natural gas from Russia, while low emission nuclear plants were closed. Consequently, the Germans spent €800 billion on measures intended to reduce emissions, but they actually went up. German lignite coal exudes 1,100g of CO_2 per kilowatt-hour compared to natural gas at 150–430g and nuclear at 16g.

Michael Shellenberger, in his 2020 book *Apocalypse Never*, described the same folly in Vermont, where climate activist Bill McKibben (350.org) and Bernie Sanders, "urged legislators in 2005 to commit to reducing emissions 25% below 1990 levels by 2012, and 50% below 1990 levels by 2028, through the use of renewables and energy efficiency… But instead of falling 25%, Vermont's emissions actually rose 16% between 1990 and 2015. Part of the reason emissions rose in Vermont is that the state closed its nuclear power plant, something McKibben advocated."

Possibly the biggest problem with nuclear is the barrier to innovation, a process which proceeds by trial and error; but who wants errors in nuclear power stations? Developments in technology evolve by trying out ideas or hunches until one works or is seen as an improvement. The outcome can then be scientifically defined and the process replicated. James Watt didn't come up with a wonderful and perfect theory for a separate condenser for Newcomen's steam engine at Glasgow University. He had a hunch on how it might work and laboured long and hard with many others to find the right materials and mechanisms. Invention is 1% inspiration and 99% perspiration. When he found that James Pickard had previously solved part of the problem and patented his idea, Watt had to find an alternative process. Patenting can be a barrier to innovation. Nuclear energy is the safest means of energy generation because it is highly regulated, but complex and inflexible regulations get in the way of nuclear innovation. Even if a particular new idea doesn't present a risk, it might take three years for the regulatory authorities to be convinced.

Westinghouse in the USA went bust when two reactors in South Carolina rose in cost from $9.8 billion to $25 billion. They were never completed but consumers still had to pick up part of the loss. The French have a long history of nuclear power, their 58 reactors providing almost 80% of their electricity, but they are old and need to be renewed. The new reactor technology, which EDF is adopting at Hinckley Point in the UK, is proving problematic for them at Flamanville in Normandy (€12 billion over budget and six years behind schedule) and Olkiluoto in Finland (up from €3 billion to €11 billion and nine years behind schedule). The French government owns EDF and cannot let it fail. Japanese technology was considered for stations at Anglesey, Oldbury (on the River Severn) and at Sellafield, but Hitachi suspended the first two because of difficulty raising the enormous funds involved. This, and the failure to follow through the 'dash for gas' power station building programme, bodes badly for future UK energy security.

Other forms of nuclear, like fusion technology, are some way off. The US company Nuscale Energy is developing small modular nuclear reactors (SMRs). They hope to test a prototype in Utah by 2025. In the UK, Rolls-Royce seeks to develop small reactors based on their designs for power units in nuclear submarines. Ten mini-reactors would produce 4,700 MW of electricity (enough for 10 million homes) and cost the same as Hinckley Point C which will have 3,260 MW of capacity.

Andrew Montford (*Reducing Emissions Without Breaking the Bank* 2020): "You replace economies of scale with economies of volume. In other words, by the time you have built your tenth module, you are very good at it, and costs fall precipitously, or that's the theory… The major risk with a nuclear reactor is overheating, and big reactors need pumps to ensure an adequate flow of cooling water around the core. They also need backup electricity supplies in case anything goes wrong (and backup to the backup as well) adding greatly to the cost and complexity. However, with a small reactor, natural convection is enough to ensure that the core doesn't overheat… Modularisation should mean much lower regulatory costs." Once the safety of one module has been demonstrated, further ones should get quicker approval.

GAS GOT CHEAPER THROUGH INNOVATION AND SET THE BAR HIGH

The contrast with developments in gas extraction technology could not be more stark. The hydraulic fracturing process evolved 'under the radar'. Matt Ridley (*How Innovation Works* 2020): "Because of mineral rights belonging to local landowners, rather than the state, and because oil companies have never been nationalised, as they were in so many other countries from Mexico to Iran, America had a competitive, pluralistic and entrepreneurial oil-drilling mindset, manifested in a 'wildcat' industry, backed by deep pockets of risk capital." George Mitchell and Nick Steinberger could experiment underground on the Barnett shale with various techniques until one was successful. Early success was threatened in 2015 by OPEC which wished to destroy this new competitor by flooding the market with cheap oil and gas. This just inspired the American oil and gas industry to innovate more, and be capable of production at as low as $40 per barrel.

There was also the threat from the green movement. At first, in 2011, Senator Tim Wirth had welcomed fracking as a stopgap measure, with relatively low emissions gas bridging the transition to renewables. But it priced renewables out of the market, despite carbon taxes and renewables subsidies. Ridley: "In the heartlands where fracking began, Texas, Louisiana, Arkansas and North Dakota, there was little opposition. A lot of empty land, a long tradition in drilling and a culture of can-do enterprise ensured that the shale revolution prospered unhindered by much if any local protest.

"But when it spread to the East Coast, to Pennsylvania and then New York, suddenly shale gas began to attract enemies, and environmentalists spotted an opportunity to fundraise on the back of opposition. Recruiting some high-profile stars, including Hollywood actors such as Mark Ruffalo and Matt Damon, the bandwagon gathered pace. Accusations of poisoned water supplies, leaking pipes, contaminated waste water, radioactivity, earthquakes and extra traffic multiplied. Just as early opponents of the railways accused trains of causing horses to abort their foals, so no charge was too absurd to level against the shale industry. As each scare was knocked on the head, a new one was raised. Yet despite millions of 'frac jobs' in thousands of wells, there were very few environmental and health problems."

The fracking revolution made America the country with the biggest reduction in CO_2 emissions, by speeding up the transition from coal to gas. Despite the great depth of the fracking process, there are occasional vibrations at ground level which the alarmists call 'earth tremours' or 'earthquakes', but which register on the Richter Scale at the same level as a bin lorry passing by, or a train rumbling over a bridge. Activists in England forced the government to pass legislation imposing impossibly small vibration levels, much smaller than for comparable industrial processes like mining, such that fracking was effectively banned. So, for essential gas supplies, the UK imports gas from the USA and Qatar, with all the additional carbon emissions from freezing the gas at source and transporting it across the oceans.

It is now time to consider how renewables compare.

HOW DOES WIND POWER PASS THE VIABILITY TESTS?

1. Wind power is unreliable and intermittent because we cannot predict when the wind will blow, and it blows at an effective speed less than half of the time in most places it is currently used. The largest onshore wind farm in the UK, at Whitelee just south of Glasgow, has a load factor (effectively efficiency) of just 27%. In other words, its utility is only present for about a quarter of the time.

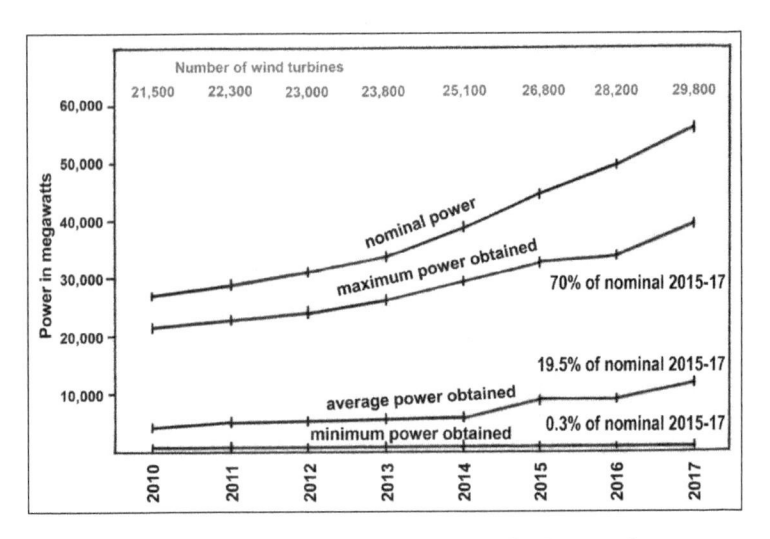

Figure 27 German wind power 2010–17. Source: VGB Powertech.

Germany has the largest number of wind turbines in Europe. Figure 27 shows that the theoretical maximum capacity of its system in 2017 was around 56,000 MW (if the wind were to blow at about 25 miles per hour all the time, everywhere), but the maximum actually achieved was less than 40,000 MW, and the mean on any particular day was less than 12,000 MW, about 21.5% of capacity. Of course there were many days when the wind did not blow and no power was generated, and some when more power was produced than required due to low demand on windy summer days. Suppliers are still paid by the government even if the electricity is not needed.

The largest offshore wind farms can achieve a load factor up to 50%, but they are much more expensive than onshore ones. Wind turbine technology has now matured, so further significant improvements in efficiency are not likely.

A renewable like wind relies on gas-fired power

stations standing by, with their turbines turning, ready to kick into action at a moment's notice, equivalent in total capacity to at least 80% of the wind power system (Rupert Steele, ex-Director of Regulations at Scottish Power estimated that 30 GW of wind generation requires 25 GW of backup, either by fossil fuels or battery storage). And of course, when these turbines are turning in readiness, they are burning fossil fuel and emitting CO_2 while producing no electricity. They need government subsidies to carry out this stupid, worthless task. We subsidise fossil fuel generation to support subsidised renewables.

Wind turbines involve the use of a lot of fossil fuel energy to extract the raw materials required, smelt the steel using coking coal to achieve the very high temperatures required (about 150 tonnes of coal per turbine), transport the materials and finished turbines around the world, install the turbines, including the concrete foundations, erect the infrastructure to take the power to where it is needed, the various maintenance activities during the 12–15-year economic life of the systems and finally to dismantle and dispose of old turbines with the landscape/seascape repaired. This must be considered when assessing net benefits.

2. The UK Climate Change Committee, along with many other activists, claim that wind power is now cheap. There is no basis for this assumption, only flawed data from the likes of the UK government's Department of Business, Energy and Industrial Strategy (BEIS) which produced its *Electricity Generation Cost*s in January 2020. However, it did not publish this so that experts and the public could question its findings. It was later forced to because of a Freedom of Information (FOI) request on

another matter made to Ofgem, whose response would refer to it. The BEIS costings were published on 24 August 2020, so that the Ofgem response on 25 August could quote them. This allowed experts like Gordon Hughes at the School of Economics at the University of Edinburgh to review it in a paper entitled *Wind Power Economics: Rhetoric and Reality*. He has data from 955 observations between 2010 and 2019 for 199 wind farms. The BEIS costs for wind power have a number of fatal flaws, adopted without question by the CCC.

The actual cost of completed offshore wind projects (at 2018 prices) rose from £125/MWh (2008–09) to £152/MWh (2018–19). While there have been economies of scale on the actual turbine blades, the big cost increases result from wind farms being in deeper water, further from shore with more costly civil engineering works, transformers and transmission plant, and other electrical equipment. The most recent UK Contracts for Difference (CfD) bids came in at as low as £45/MWh, comparable with the cost of gas generation. Hey presto, there was a 70% drop in price in the course of a few years, without an obvious reason. The government swallowed it, hook, line and sinker, and immediately announced a massive increase in offshore construction. The loss-leader tactic of the bidders had worked, and they will no doubt charge what they like when too much of our energy is in this basket and our energy security is threatened.

Hughes: "The offshore wind sector is dominated by large, often state-controlled companies that can deploy large cash flows from existing generation and/or network businesses, and which are under little pressure to return cash to either their customers or their shareholders. Operators may expect to be able to sell on a large portion of the shares in their projects to over-optimistic investors with little appreciation of the risks involved." In other words, the public subsidies and high energy prices have made foreign, state-owned companies like Vattenhall of Sweden so rich that they can afford to take big investment risks with their plentiful spare cash, and sell stakes to opportunity-hungry, green-leaning private investors who may not think a predominantly public sector company/employer would be allowed to go bust. Public sector players seldom have a beneficial effect on market forces, and these investors should beware since about 40% of all offshore projects exceed their anticipated costs by more than 25%.

The proof is in the pudding. There is a direct correlation between the amount of renewables European countries have and the cost of electricity (Figure 28 – data compiled by Euan Mearns from

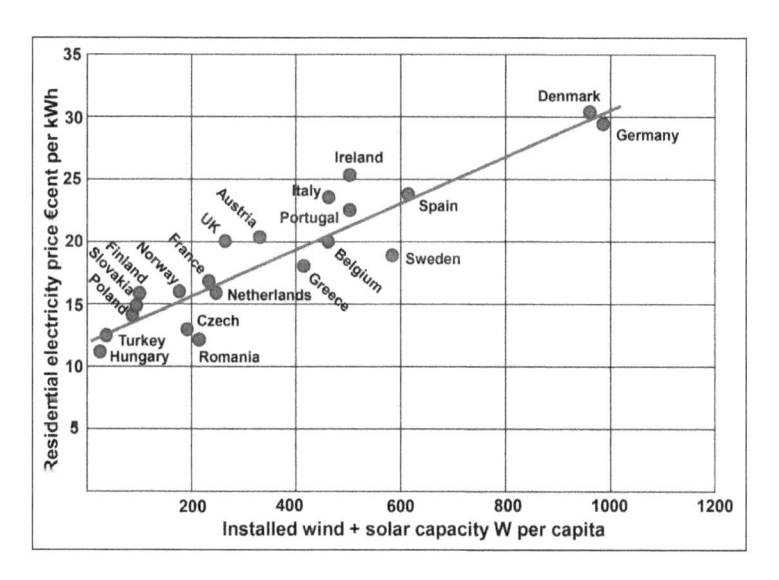

Figure 28 Europe residential electricity price and installed wind and solar capacity. Note: the commercial electricity price is proportionately higher in the UK than in many parts of Europe because of an £18/tCO$_2$ carbon tax which the CCC wishes to raise to £75/tCO$_2$.

BP2015 and Eurostat sources). It is not just the costs of the wind farms but the massive infrastructure to gather the electricity from diverse locations and direct it to where it is needed. There is also the cost of grid balancing due to the intermittent nature of the wind, the costs to the National Grid approaching £2 billion in 2020 from the £367 million spent in 2002 before the growth in renewables.

Accounts in 2020 from energy companies, the first from the Moray East wind farm, confirm that the actual cost of offshore is as high as predicted. Andrew Montford from Net Zero Watch: "Except for the wind lobby, there is now widespread agreement that Contracts for Difference auctions do not reflect underlying costs. The hard data from audited accounts is now giving backing to this reality. It is clear that offshore wind is extremely expensive, and will remain so for the foreseeable future." Craig Mackinlay MP: "Boris Johnson assured me that the cost of offshore wind has fallen by 70%. Sober analysis shows beyond all reasonable doubt that this is not the case. Not only does this show that the PM is being given flawed information by his advisers, the public is being led into a cost and energy security disaster."

3. Wind farms cause harm to the environment because they take up so much space, and they kill birds and bats. The London Array 175-turbine wind farm in the Thames Estuary has a site area of 122 km², with about half of the average daily output of a gas-powered station that could be fitted on less than a square kilometre. There are 6,260 turbines in Denmark such that it is possible to walk from one end of the country to the other within sight of them all the time, when the 1.5 GW capacity could be replaced by a gas plant. The countryside has been

industrialised by wind 'farms'. A study carried out by the John Muir Trust in 2018 showed that wind turbines can be seen from almost half of the Scottish mainland. Do we want our beautiful countryside to continue to be industrialised by a plethora of spinning (or not spinning most of the time) white metal objects?

Wind farms chop up birds, especially raptors and seabirds on migratory coastal sites. Bats evade the 300 km/h spinning blade tips, but are suffocated (barotrauma) by the pressure drop behind the turbines (YouTube *Bird v Wind Turbine*, and *Wind Turbines v Birds and Bats*). A 2015 study of 12,841 wind turbines in North Germany estimated that they killed 7,865 buzzards, 10,370 ringed pigeons, 11,843 mallard ducks and 11,197 gulls. The German Wildlife Foundation called for a ban on new wind turbines close to forests until further studies could be conducted.

Michael Shellenberger quoted various German studies on insect kill by wind turbines. "In 2001, researchers found that the build-up of dead insects on wind turbine blades can reduce the electricity they generate by 50%... [with] a loss of about 1.2 trillion insects of different species per year... The impact of German wind farms is 'not limited to local populations, but includes species like the ladybird beetle (C. septempunctata) and the painted lady butterfly (V. cardui) that travels hundreds and even thousands of kilometers through Europe and Africa.'... While much of the media coverage has blamed industrial agriculture [for insect numbers decline], it is notable that the biggest insect population declines are being reported in Europe and the United States where the land dedicated to agriculture has declined over the last two decades. What have spread are wind turbines." Much more research is required.

CONCLUSIONS ON WIND POWER

Those developing wind turbines over the last 40 years have been vainly trying to squeeze efficiency out of a 13th-century technology. Wind power is very inefficient because of the vagaries of the weather and the physical laws of Newton or, to be more specific, Betz's Law, formulated by Albert Betz in 1919, which concluded that no turbine can capture more than 16/27 (59.3%) of the kinetic energy of wind. We have known for a hundred years the limits of wind power efficiency and that it is not a sensible option for modern mass energy generation. The Victorians who substituted steam-powered ships for sail would question our sanity. Perhaps Boris Johnston was correct when he said a few years ago that wind turbines would not blow the skin off a rice pudding. With the CCC's 'backbone' of offshore wind power we are sure to flounder on the rocks of reality.

The UK's energy policy involves taking about £10 billion each year away from vital public services, to give in subsidies to rich international renewables suppliers, mostly to provide wind power, which makes the country's energy costs so high that we have to export our industry. Of course, the emissions transferred to places like India and China are merely brought back to the UK, embedded in manufactured goods. The 42% reduction in domestic emissions since 1990 can be almost completely accounted for by the move from coal to gas for power generation, the deindustrialisation of the country, and the assumption that biofuels are carbon neutral, which they are not. The benefits of wind power are barely significant. The current policy makes no financial sense, and one might say that it is morally reprehensible because of the fuel poverty it exacerbates.

On the first day of COP26, Scottish National Party First Minister Nicola Sturgeon boasted that Scotland had so many wind turbines that 97% of its electricity can be generated by renewables. The next day the wind dropped and the overnight temperature fell to 3°C. It does not matter how many wind farms are built, when the wind does not blow they do not generate electricity. The website energynumbers.info tells us where the UK's electricity comes from on any particular day.

Gas turbines had been turning at significant cost, without generating electricity but producing emissions, waiting to kick in when the wind dropped, but this was not enough to keep the lights on. Solar power provides nothing to the grid at that time of the year with a low solar elevation and short days. The next option was the expensive interconnectors the UK has built to continental Europe, but they were also having problems that Tuesday so more electricity went out of the UK's grid than came in. Interconnectors involve a tug-of-war such that the country with the wonkiest power system at the time pays more to secure the electricity. So, to keep the lights on, the two coal-fired generators at the Drax Power Station in Yorkshire were switched on. Intermittent coal power is expensive, resulting in a cost of £4,000/MWh, against the normal costs of coal or gas generation of about £40/MWh. With energy costs soaring, and despite a warming world, many Scots cannot afford to heat their homes in winter.

Discussion: Why are all wind turbines white, rather than a colour such as green which might make them blend into the landscape better?

HOW DOES SOLAR POWER PASS THE VIABILITY TESTS?
1. Solar power is unreliable and intermittent because we cannot predict when the sun will shine, and even in countries where it does shine a lot, it doesn't at night, that is, half the time. In northern Europe, northern USA, Canada, Russia, and most of Chile, Argentina and New Zealand, solar power is very limited in winter (when demand is high) due to long nights and low, hazy sunlight, with less than 10% efficiency. An Adam Smith Institute report by Capell Aris, *Solar Power in Britain*, noted that Britain's 32 million solar panels operate at more than half of their capacity for only 210 hours per year (2.4% of the time).

    A rough calculation suggests that the energy

used sourcing raw materials, manufacturing the panels, installing them and their grid infrastructure, maintaining the system and disposing of the panels at the end of their useful life (perhaps 20–25 years) is more than the energy actually generated from them when installed beyond the latitudes of 35°N and 35°S. Such a generalisation may or may not be useful but, instead of just regarding each solar installation as 'virtuous' and awarding it a public subsidy, a proper business case should be made for each solar farm and housing development with roof panels, proving value for money, taking into account direct and indirect costs and benefits, and actually demonstrating that emissions are saved. Like subsidies for expensive electric cars, home solar panels tend to benefit better-off people more who can afford a new home.

In sunnier climes, for example in deserts, a theoretical efficiency is $20W/m^2$, but in practice the energy produced is between $5W/m^2$ and $14W/m^2$. Keeping dust off glass solar panels is tricky. A desert storm in coastal areas like the Gulf States mixes water with sand and coats buildings and solar panels in mud. People do not generally live in deserts so there are huge inefficiencies transmitting the power over long distances to where it is needed.

Like wind power, solar relies on gas generation backup, although it is useful to supplement power for air conditioning during the day when temperatures are hot in tropical countries. Small domestic units are also useful for providing hot water in sunny countries and, if you have a food refrigeration plant that has high energy demand on warm sunny days, solar might be for you.

2. As we saw above, solar is as expensive as onshore wind generation. The technology is not fully mature,

so some more efficiencies may yet appear, but these are unlikely to be large.

3. Even in deserts there is wildlife, often with species which are precariously hanging on to existence. Do we want to cover hundreds of thousands of square kilometres of desert with glass? Birds can confuse them for lakes, land on them and be burned. The change in surface albedo (reflection characteristics) from sand to glass in the largest farms also has a significant effect on the local climate.

Solar photovoltaic modules use at least one rare or precious metal like silver, titanium dioxide, calcium telluride and copper indium gallium selenide, mined in various parts of the world, with inevitable environmental damage due to the large quantities of excavated spoil. A large amount of fossil fuel is used in the mining and refining processes.

STRATEGIES TO REDUCE THE INEFFICIENCY OF WIND AND SOLAR – INTERCONNECTORS

To try to counter the inefficiency of renewables, many countries build undersea or overland cables so that when a particular region has a power demand beyond the capacity of its own renewables, it can import it from another country. This is a flawed strategy for a number of reasons. There are the capital costs and transmission inefficiencies in such a long-distance transfer of energy, assuming that the wind turbines are working in areas that the cables reach, and they are not also becalmed. When we obtain electricity from another country, we do not necessarily know how it has been generated, so Britain may be tapping into continental electricity which partly comes from German coal-powered generators, with a high embedded emissions content. Undersea power cables are much more expensive and problematic than overland ones, but they are out of sight.

At the time of the 'Beast from the East' cold snap, in

March 2018, Tony Lodge writing in *The Daily Telegraph* noted that Britain would rely on a fifth of its electricity supplies by 2025 from undersea interconnectors. These "respond to price signals and electricity flows through them to where the price is strongest. Consequently, if there is a sustained period of little or no wind and sunshine across the rest of Europe, coupled with high demand, then the UK cannot rely on imports to keep the lights on. In fact, interconnectors could exacerbate a supply shortage as power would be exported out of the UK to the continent when prices are higher there." Most household heating is electric in France from nuclear plants which cannot be turned up at short notice, so it is more likely to import energy to meet peaks. "It is estimated that Britain would need 26 GW of additional gas generation capacity by 2030 to plug any potential gap left by cloudy, windless days and to replace the electricity output from closing older coal, oil and nuclear plants. On current trends, however, the UK is on track to build just 12 GW by 2030. This goes some way to explain the hurry to build interconnectors."

However, there is no guarantee that the power will be available when required at the ends of these connections, or that the UK's old partners in the EU would be willing to supply it. In November 2020, when the Brexit negotiations were still at their height, President Macron threatened to hold the UK to ransom by withholding electricity from his nuclear plants unless the UK gave him more access to its fishing grounds. The next negotiation takes place in 2026, when the UK may be experiencing power cuts or load sharing due to over-reliance on wind power, and when perhaps France will have more success.

The study shown in Figure 27 also noted that, on the calmest days of 2014, wind power only achieved 0.075% of nominal production, and if Germany had been linked up to seven neighbouring countries, this figure would have barely reached 1% of demand. Unwindy anticyclones affecting Germany usually extend to neighbouring countries, so interconnectors to them would be useless.

Discussion: What are the effects of power cuts on individual people and businesses, which happen when the grid cannot be adjusted quickly enough to meet peak demand for electricity? What can the individual do to keep their home or business powered should the mains supply be cut off?

STRATEGIES TO REDUCE THE INEFFICIENCY OF WIND AND SOLAR – GRID BATTERY STORAGE

We should not have built wind and solar farms until we had worked out a solution to their intermittency. Current battery technology is a poor substitute for fossil fuels. A March 2019 report by Mark P Mills of the Manhattan Institute showed that 100 barrels of oil are needed to fabricate a quantity of batteries to store a single barrel of oil-equivalent energy. "Two hundred thousand dollars' worth of Tesla batteries, which collectively weigh over 20,000 pounds, are needed to store the equivalent of one barrel of oil… The annual output of Tesla's Gigafactory, the world's largest battery factory, stores three minutes' worth of annual US electricity demand. It would require 1,000 years of production to make enough batteries for two days of US electricity demand."

Donald Sadoway, at the Massachusetts Institute of Technology (MIT) does not see existing battery technology being able to be scaled up to fit the bill. He and one of his sponsors, Bill Gates, believe that "The fundamental approach to grid level storage ought to be intrinsically different from the approach to mobile storage." Starting in 2006, his team came up with a Liquid Metal Battery (LMB) rather than a solid state one, with liquid magnesium (light density) and liquid antimony (high density) as the electrodes, and a molten salt electrolyte between. The French energy company Total is his other sponsor, but the project would be moving faster if they believed it was definitely the answer to the problem. The point is that people are investigating new solutions, and it seems likely that some will prove to be viable and affordable at some point in the future; but they are not just now, so we cannot count on them in any plan we make.

Bill Gates in his 2021 book *How to Avoid a Climate Disaster: The Solutions We Have and the Breakthroughs We Need*, starts with the usual assumption that more emissions will have a big effect on temperature, and that "something like a fifth of carbon dioxide will still be there in 10,000 years", then puts his faith in new technologies which we currently do not have and do not know whether we will be able to develop effectively: hydrogen produced without emitting carbon, grid-scale electricity storage that can last a full season, electrofuels, advanced biofuels, zero-carbon cement, zero-carbon steel, plant- and cell-based meat and dairy, zero-carbon fertiliser, next-generation nuclear fission, nuclear fusion, carbon capture (both direct air capture and point capture), zero-carbon alternatives to palm oil, etc. There is complete faith that new technologies will solve all our problems, except perhaps we should better define the problem first to prove that it really exists.

STRATEGIES TO REDUCE THE INEFFICIENCY OF WIND AND SOLAR – HYDROGEN PRODUCTION

Hydrogen is a flammable gas and when burned produces water as a by-product ($4H + O_2 > 2H_2O$) instead of carbon dioxide and water when methane gas is burned ($CH_4 + 2O_2 > CO_2 + 2H_2O$). Water vapour is not regarded as so powerful a greenhouse gas as carbon dioxide, partly because it is believed to stay in the atmosphere for less time. Some people reckon that we should use hydrogen where we now use methane and petrol, in other words in homes and in vehicles. Our natural gas cookers and boilers would be powered by hydrogen, although we would have to buy new ones that worked with the different gas.

James Lovelock: "[Hydrogen] tends to make steel brittle, and because of its small molecular size it leaks easily through tiny holes that would be only a minor problem with a heavier gas such as propane. A hydrogen-air mixture detonates when ignited, instead of burning fast but smoothly as does a methane-air mixture. A hydrogen flame is invisible, so that the ignition of a small leak can cause dangerous overheating of piping or valves before it is noticed. All of these drawbacks can be overcome by good engineering, but they add to the time and cost of establishing a hydrogen economy."

We have to use lots of fossil fuel to break down water or methane to produce the hydrogen in the first place. This defeats the purpose of the whole thing. *Missions Possible: Reaching Net-Zero Carbon Emissions from Harder-to-Abate Sectors by Mid-Century*, by a group called the Energy Transitions Commission, reckons that the cost would have to reduce from $1,000 per kW today to about $250 per kW to be viable.

However, in some countries there are now so many wind turbines that they produce more energy than is required on windy summer days when electricity demand is low. Perhaps we could use this excess power to produce hydrogen, effectively a way of storing wind energy in a different form. Those who advocate this believe that wind energy is the cheapest way of producing energy, which we have already seen is false. The building of many more wind farms to produce lots of excess power in summer would be expensive and cause further environmental harm. Chris Stark, chief executive of the UK Climate Change Committee, recognises that we could not build enough wind farms to meet the demand for hydrogen that his team envisages. Hydrogen as an affordable and plentiful fuel does not seem viable, apart perhaps for its limited use in urban transport to reduce the pollution from diesel engines, if these cannot be made more clean.

CARBON CAPTURE AND STORAGE

So if biofuels are environmentally unacceptable and wind power is intermittent, we are left with natural gas as our main way of generating electricity. If only we could capture the 'nasty' CO_2 at power stations and pump it back into the voids left by our North Sea gas and oil extraction. The technology can work. The Carbon Capture, Utilisation and Storage (CCUS) Association cites the Quest facility in Alberta, Canada, which captures a million tonnes of CO_2 each year from a hydrogen-producing facility, and injects it into voids 2 km below the surface. The Sleipner project in Norway is the oldest offshore CCS facility, being operational since 1996, also with a million tonnes of

CO_2 captured each year. Early in its life it had to stabilise the sands around the injection site to stop leaks, a process that was successful. Despite having shared data with the research community for 15 years, it does not appear to have encouraged the development of similar projects. Operating costs are $17/tonne of CO_2 (tCO_2), including carbon tax relief and carbon credit from the EU's emissions trading scheme.

London-based Carbon Clean is developing modular carbon capture units for use in industry, aiming to reduce the currently expensive $70/$tCO_2$ systems down to $30/$tCO_2$, and with an efficiency of 80%. It claims that its 2016 purpose-built plant near Chennai in India is the world's first fully carbon capture, storage and utilisation plant, with an efficiency of 90%. Their process uses solvents, like monoethanolamine, to remove the CO_2, which must be safely disposed of. The UK doubled the carbon tax to £18/tCO_2 in 2015, so it would need to more than double again to make such CCS competitive, but CCC is proposing "a £75/tCO_2 charge across most domestic sectors" (*Sixth Carbon Budget Policy* page 28) to raise £27 billion each year to fund its masterplan.

The main reason given for the lack of investment in this technology is that trapped CO_2 is effectively worthless, so no commercial company is interested in developing it. There is a limit to how much CO_2 is needed for fizzy drinks and other food processing. Projects would therefore be wholly government funded, which would take investment away from other, more obviously attractive public projects like new schools and hospitals. The UK government rejected the funding of a trial project in 2011. A lot of money could be spent before we are sure that the saline aquifers and voids at trial sites are secure storage locations, especially for the enormous quantities of CO_2 requiring to be disposed of.

WHY TRANSPORTATION IS THE BIGGEST CHALLENGE TO EMISSIONS REDUCTION

The best energy storage technology we have just now for electric cars is lithium ion batteries. But lithium is a rare and expensive element, and is best suited for small, mobile devices like phones and computers. Compared with gas or petrol, it is very poor at storing electricity. A kilogram weight of petrol stores 60 times more energy than the same weight of lithium battery (46,000 kJ/kg against 720 kJ/kg).

Range anxiety is the fear that the battery will run out of power before the journey is completed. Horiba Mira, an engineering consultancy, noted that manufacturer battery range tests are done in ideal conditions, usually in an air temperature of 23°C, without heating or air conditioning on in the car, and without a full passenger and luggage load. Advertised figures can overestimate the range of the car by up to a third. For example, a 2016 Nissan Leaf 30 kWh car had an advertised range of 155 miles. In a test in Britain, it never went higher than 120 miles when fully charged, and the range was as low as 86 miles. This is a big disincentive to buying an EV, so people tend to buy hybrids, which is a complete cop-out.

In 2015 Tesla, owned by the richest man in the world Elon Musk, stated that it would need to use all the world's current supplies of lithium to meet its target of half a million EVs by the end of the decade (it nearly got to 400,000 by 2020 although the company was still loss making at the beginning of that year, despite the Obama administration awarding it subsidies of nearly half a billion dollars). Consequently, according to Benchmark Minerals, the cost of lithium hydroxide almost trebled during 2016. The market usually responds and should find new sources of lithium, but this may take many years.

There are other minerals involved. According to Glencore, a shift of 10% towards EVs would require an extra 400,000 tonnes of nickel, which is used in electrodes. The total world supply is currently 2 million tonnes. Cobalt can also be used for electrodes, but this material is in high demand for smart phones and tablet computers. The conclusion is that lithium ion batteries are unlikely to drop in price from their high level because the component materials are rare (electric cars tend to be about £10,000 more expensive than their petrol equivalents), they rely on a lot of fossil fuels to produce, and the mining of rare minerals will cause a lot of environmental damage.

Petrol in cars would have to be substituted by electricity from charging points, with many extra gas-fired power stations capable of meeting the surges at early evening and early morning peak charging times, and an upgraded electricity distribution system to carry this increased charge to all homes and businesses. As Tony Lodge has already pointed out, the UK will be 14 GW short of gas power station capacity by 2030, without extra power for charging EVs. The time, cost, labour supply and materials production involved in the gargantuan effort that would be necessary to resolve this would take decades and use a huge extra amount of fossil fuel energy. It makes no sense to set a date for substituting EVs for petrol and diesel cars, even if it were thought beneficial to have them at all.

Volkswagon reckons its eGolf might need to drive 50,000 miles before the fossil fuel energy used to produce the metals, glass, plastics and batteries in each car is expunged. By this time the battery will probably need to be replaced, with more emissions added. This of course assumes that the car is being charged with zero-carbon electricity which we know it will not be. In other words, each EV has little chance of ever reaching a carbon neutral state. Heavy goods vehicles, boats and planes require too much power to rely on pathetically inefficient lithium batteries. Hydrogen may have a very limited use as mentioned above but, as one US senator said: "I would not like to fly in a solar-powered plane at night."

WHY NOT USE LESS ENERGY?

We can better insulate our homes and take a bit more care on how we use energy, but the percentage savings would be tiny. The Centre for Alternative Technology's 2013 (updated 2019) *Zero Carbon Britain: Rising to the Climate Emergency* (Greta Thunberg's mentor Kevin Anderson wrote the foreword) envisages a drop in heating demand of the average UK home from 10,000 kWh/yr to 4,000 kWh/yr by improving insulation and reducing air leakage, but also by having people (forcing people to?) live in colder internal temperatures. But there is no such thing as an average house. Older houses are difficult to upgrade and most modern houses have been built with high insulation standards such that it would be difficult to make significant improvements. And who is going to tell old people to turn their heating down in winter? The authors seem to envisage that everyone will embrace their spartan, puritanical cause and be happy to suffer for the planet. Why, if the world is becoming warmer, do we need to experience more cold in our homes?

The strategy also envisages that we just stop moving about so much (a permanent COVID lockdown?) unless we walk or cycle more because it is difficult to make much of a saving in the various means of mechanised transportation. It reckons that demand would have to reduce from the present UK 687 TWh/yr (terawatt-hour per year) to 380 TWh/yr, a 45% reduction in travelling, with more efficient means of transport (walking, bikes, trains) taking this down to 154 TWh/yr. In other words, the energy used in transportation would have to drop by nearly 80%. How will people be compelled to do this?

Most of our energy in a modern lifestyle is expended outside the home. Should we ask hospitals to save 20% on their energy requirements? Should they stop using energy-hungry, high-tech equipment and have fewer ambulances? Should the police force cut back, with fewer cars and high-tech, energy-sapping forensic processes? Should schools and colleges turn down their heating in winter? Should we give up our mobile phones and stop texting each other, since every message requires energy, and the cumulative effect of billions of people doing so at the same time is enormous? Should we stop subscribing to so many television channels and cut down the number of international sports events covered, with the energy used by spectators, TV production crews and the mass communications technology? What about cutting the increasing production of drama in film and TV to take away the livelihood and carbon emissions of actors like Leonardo DiCaprio, Mark Rylance, Jude Law, Emma Thompson and Benedict Cumberbatch who support Extinction Rebellion and fly round the world to preach to us about climate change?

Even if we made all these sacrifices in rich countries, they would be tiny compared to the increases happening in the rest of the world. Supposing all the half billion people in Europe decided to take one for the planet and commit suicide. The 3.5 gigatons of CO_2 emissions saved each year would be cancelled out by the 4 gigatons of extra emissions each year allowed to be produced by India and China by 2030 under the 2015 Paris Agreement. And of course these countries are still building many coal-fired power stations which they will not close down prematurely in 2030 to start reducing their own emissions. As I said at the beginning of this chapter, it is certain that global emissions will rise to 2040 and beyond. So we need to plan for this and take whatever steps are thought appropriate to adjust to the consequences.

> **Discussion:** What means could you adopt to significantly reduce the amount of power you use? Could you walk to school or college? Could you manage without a mobile phone or a family car? How would this affect your lifestyle? What about your older, less mobile family members (about 90% of visits to hospital are by car or ambulance in the UK)?

THE COST AND VALUE FOR MONEY OF NET-ZERO POLICIES

Politicians have a habit of rushing into energy policies without estimating the costs or doing a cost/benefit analysis. Since much of the policy relies on technologies that are currently unproven, and therefore financially unpredictable, this is understandable.

Bjorn Lomborg (*False Alarm: How Climate Change Panic Costs Us Trillions, Hurts the Poor, and Fails to Fix the Planet* 2020) has endeavoured to explore the costs of energy policies and their impact on the climate. This requires making two assumptions: that policies will actually achieve net-zero emissions by the set date, even though I have shown that the technology doesn't yet exist to achieve this; and that the UN estimates of the transient climate response are valid, which I have also shown are highly dubious. He found that New Zealand was the only country honest enough to publish figures that he could work on.

"In 2007, Prime Minister Helen Clark declared her vision for the small nation to become carbon neutral by 2020. She was celebrated by the United Nations as a 'Champion of the Earth'. If only cutting carbon was as simple as winning attention. New Zealand not only failed to achieve the vision, but also failed to reduce any emissions. The latest 2019 official statistics show that the country's total emissions will be higher in 2020 than they were when Ms Clark's ambition was declared... Yet, in 2018, Prime Minister Jacinda Ardern renewed the pledge, promising to achieve carbon neutrality by 2050.

"To its credit, Ardern's government actually asked its leading economic authority to estimate the cost of her promise. Thus, we have what is likely the only official, academically credible estimate of what it will cost to achieve carbon neutrality. This research shows that just getting halfway to the target, cutting 50% of New Zealand's emissions by 2050, would cost at least $19 billion annually by 2050... getting all the way will likely amount to more than $61 billion annually, or 16% of GDP by 2050... And even the 16% GDP cost relies on a fairy tale assumption that every single policy will be enacted as efficiently as possible. Bearing in mind the evidence that costs double in the real world, it could be 32% or more... The cost would be the equivalent of at least $12,800 for every single New Zealander, every year. If the policies are done badly, as they have been done so far across the globe, the cost per person could even go beyond $25,000 per person.

"If we assume that New Zealand this time will actually deliver on its net-zero promise in 2050, and stick to it throughout the rest of the century, the total amount of greenhouse gas reduction will, according to the standard estimate from the UN's climate panel, deliver a temperature reduction in the year 2100 of 0.004°F. Given the expected temperature increase by around 2100, this means that New Zealand going net-zero by 2050 will postpone the warming that we expected to see on 1 January 2100 by about three weeks to 23 January 2100. So, New Zealand is considering spending at least $5 trillion to deliver an impact by the end of the century that will be physically unmeasurable... Sooner or later, and likely sooner, a politician is successfully going to argue to dump the net-zero promise that will deliver zilch in a century, and instead double spending on things like health, education and the environment, and get some tax reductions.

"The Paris Agreement will be the costliest pact ever agreed to, by far. It will cost us $1–$2 trillion per year from 2030 onwards, if actually fully implemented. Yet the agreement will do almost nothing for the climate: all of its promises will reduce the temperature rise by the end of the century by an almost imperceptible 0.05°F. And none of the big emitting countries are anywhere near close to actually delivering on their promises. Spending trillions to achieve almost nothing is, not surprisingly, a bad idea. Every dollar spent will produce climate benefits worth just 11¢."

Imagine you are a financial advisor, like Stuart Kirk of HSBC. You and your colleagues know that the cost to your clients of climate change over the term of their investments, or even for the rest of the century based on UN assumptions of global warming, will be tiny compared to the increased value expected from their funds, so small that it is not worth bothering about. It is your professional and legal responsibility to advise them of this, but if your organisation is accused of not taking climate change seriously enough, there could be corporate reputational damage. What Kirk did was deliver a candid talk at a *Financial Times* conference, with the knowledge of his colleagues, followed by his superiors suspending him, subject to investigation. The bank could state that its official policy did not accord with Kirk's analysis, which is available on YouTube, *HSBC's Stuart Kirk tells FT investors they need not worry about climate risk.*

The titles of presentation slides give an indication of his narrative.

- Unsubstantiated, shrill, partisan, self-serving, apocalyptic warnings are ALWAYS wrong.
- As [past] warnings became ever graver, the more asset prices INCREASED in value.
- Even by the UN IPCC's own numbers, climate change will have a negligible effect on the world economy.

- Adaptation is cheap and effective.
- Climate-related costs, relative to GDP and mortality rates, are down.
- Even if climate risk isn't negligible, it's too far into the future to matter for most companies.
- To make climate change appear like a significant threat, scaremongers are torturing their models (it is easy to show that climate change is an investment risk if you engineer a parallel bond market collapse, as the Bank of England does in some of its scenarios, and it also uses the IPCC's impossibly gloomy RCP8.5 scenario).

In pandering to clients who have bought into the UN's narrative, frightened to point out they are wrong, the heads of leading financial institutions are effectively in collaboration with anti-capitalist eco-anarchists like XR founder Roger Hallam. Ben Pile, in a YouTube video *Intended consequences: energy price rises and inflation*, lists various unelected financial patriarchs who align with UN COP strategies, with Michael Bloomberg at the head of the Taskforce for Climate-related Financial Disclosures, a modern version of the Sierra Club, and Mark Carney, former governor of the Bank of England, UN special envoy on climate change, founder of TCFD and head of the FSB.

From 2006 to 2009 Rishi Sunak was a director in Sir Christopher Hohn's TCI company which Pile claims funds through a philanthropic organisation CIFF, directly and indirectly, organisations like XR, The Green Alliance, Carbon Brief and UK Climate Assembly, and makes a fortune in green hedge fund activities. As Boris Johnson's UK Chancellor of the Exchequer, Sunak's policies seemed diametrically opposite to the tradition of his political party, printing money to get us through COVID-19 (fair enough) but compounding this by printing money to very slightly and briefly ease the pain of self-harming energy policy, as well as applying arbitrary windfall taxes on energy companies which are guaranteed to deter investment in fossil fuels, the only solution to the energy crisis.

The inevitable rise in inflation hurts the electorate, but those listed by Ben Pile are least affected. "The absence of cheap energy has caused the very same problems that climate change activists said would be caused by global warming and climate change: the loss of agricultural productivity, the return of hunger and poverty, and exposure to temperature extremes." Adding to this the folly of adopting 'green' agricultural policies, the Sri Lankan economy collapsed.

KEEP ALL FOSSIL FUELS IN THE GROUND?

Green ideology rejects the fossil fuels which power our comfortable, wealthy, safe and happy lifestyles, even if we have no viable and affordable alternatives. We have already seen that only coking coal gets furnace temperatures high enough to make steel and glass for wind turbines and solar panels. Plastics are also demonised. Plastics are produced from petrochemicals and are such an integral part of modern life that we take them for granted. They are unique for their combination of strength, flexibility, durability, lightness, hygiene and economy, even if they are sometimes difficult to recycle (so we have to incinerate a lot of plastic waste).

Poor countries have a poor track record in stopping discarded plastics escaping into watercourses and the ocean, but they need them most as they develop their economies. Some stupid people in rich countries are also not good at disposing of used plastics, but that problem is tiny in comparison. Once again, more economic progress and funding to build efficient waste disposal and recycling plants is the answer, not the problem.

Alex Epstein: "While you think of oil in your car as in the gas tank, in fact there is more oil in the materials in the car than in the gas tank. The rubber tires are made of oil, the paint and waterproofing are made of oil, the plastic, dent-resistant bumper is made of oil, the stuffing inside the seats is made of oil, and in most cars, the entire interior is one form of oil fabric or synthetic material or another – because oil is such a cheap and effective way to make things. When a policeman has his life saved by a bulletproof vest, when a firefighter has his life saved by a fire-proof jacket, that is oil; that is something that was once a useless raw material, now made into a resource… Oil is coveted as the world's most versatile raw material for making synthetic materials. You are probably sitting in a room with at least fifty things derived from oil, from the insulation in your wall to the carpet under your feet, to the laminate on your table, to the screen on your computer. Oil is everywhere. That is how the average American uses 2.5 gallons each day."

And yet climate activists campaign to keep all fossil fuels in the ground. Consider the huge number of products that we would miss out on if we did not drill for oil and gas.

- Antiseptics, aspirin, refrigerants, antihistamines, vitamin capsules, cortisone, dentures, eyeglasses, heart valves, bandages, artificial limbs, contact lenses, hearing aids, petroleum jelly, sterile gloves.
- Fertilisers, pesticides, herbicides, fruit nets to keep birds off, polytunnels to protect tender crops from late frosts and extend the growing season, detergents, photographic film, antifreeze.
- Electrical tape, plumbing washers and seals, water pipes, uPVC windows, paint brushes, adhesives, paints, grease for bearings, putty, rubber cement, dishwashing liquid, floor wax, refrigerator linings.
- Surf boards, footballs, basketballs, fishing rods, lines and nets, golf equipment, tennis rackets, roller skates, skis, guitar strings, tents, parachutes, awnings, balloons, artificial turf, garden hose.
- Bicycle tyres, fan belts, road markings, cats' eyes, cycle and other safety helmets, umbrellas, car battery cases, carrier bags, ice cube trays, telephone cases, shower curtains, cameras, credit cards, wellington boots.
- Nail polish, perfumes, lipstick, soap, shampoo, toothpaste, toothbrushes, hair colouring, shoe polish, hand lotion, candles, combs, condoms, hair curlers, panty hose, shaving cream.

Discussion: While we can substitute paper drinking straws for plastic ones (but they are not very good because they quickly get soggy), how many other products in the above list have alternatives, or may have done before plastics were invented? Why are the plastic versions preferred?

UK GOVERNMENT POLICY BASED ON THE CLIMATE CHANGE COMMITTEE'S RECOMMENDATIONS

I will finish this chapter by highlighting how unrealistic and naïve (green?) the CCC's *Sixth Carbon Budget* of 9 December 2020 is, which the government seems keen to adopt without question. The CCC claims to be an 'independent' quango, but the political environment it operates in, and the nature of its eight-person committee, makes this impossible. The chair is John Gummer, a politician who is also a director in an environmental consultancy with his son. There are six academics with, according to published CVs, only ten years' experience of the private sector between them. Their understanding of the consequences of their recommendations in the real commercial world must be very limited. Their theoretical, rather than practical, mindsets may also explain why they recommend various technologies which have not yet been shown to be viable or affordable. Two of them have had key roles with the UN IPCC. Only one CCC member works for a private sector company, Drax, which majors in biofuel electricity production, an energy source which this book and Greenpeace condemn for its very harmful effects on the natural environment (and reliance on fossil fuels to obtain the woodchips), but which the UK government rewards with more than three-quarters of a billion pounds in subsidies each year for its Yorkshire plant. How 'independent' is that? If there ever were a more perfect formula for groupthink, it is this.

The Committee exerts power without responsibility. Unlike private sector risk takers trying to implement its strategies, none of its members will suffer bankruptcy if their best laid schemes 'gang aft agley', and if their advice turns out to be unrealistic, unaffordable and ineffectual in terms of reducing human emissions, which my review concludes it is. The CCC can have no idea whether its radical and speculative ideas will work, but it is quite happy to use the British economy and its people in its experiment.

Its 'backbone' is wind power, especially offshore because there are few suitable locations left onshore and, of course, it is 'cheap'. To address the problem of intermittency (it only mentions this awkward word seven times in its 997 pages), it has backup natural gas and biofuel power stations which would be fitted with carbon capture and storage. Millions of homes would have heat pumps installed, and those that didn't would use electric or hydrogen power, as would businesses. All large vehicles like buses and goods vehicles would run on hydrogen. Biofuel burning is environmentally unacceptable, and if we need almost 100% gas backup for wind turbines when there is no wind, why do we not just omit wind power and get all our electricity generation from gas burning? That would mean natural gas generation with CCS as the 'backbone', and no one has the slightest idea if CCS is viable at grid scale, or what the costs might soar to.

We would need more gas-generated power for wind backup, to replace the petrol and diesel in EVs, to provide electricity for heat pumps and to manufacture hydrogen, never mind the extra power needed to effect this huge infrastructure transformation. That is, to apply low-carbon technologies we need lots more carbon-based fuel on an ongoing basis. Where do we get this methane from since North Sea reserves are dwindling? Do we build pipelines from Putin's Russia via the Baltic the way Germany has done, or do we cast off our irrational loathing of fracked gas and tap into the supplies that are beneath our feet?

The Report's authors have no experience of procurement processes. They major on the 'what to do' but seem to have no conception of the 'how to do it'. It is a wish list with an artificially notional timetable, driven by the panic of an alleged 'climate emergency'. Even if the private sector can instantly deliver the innovation necessary to roll out the proposed technologies that are still in their infancy, the demand on raw materials, products,

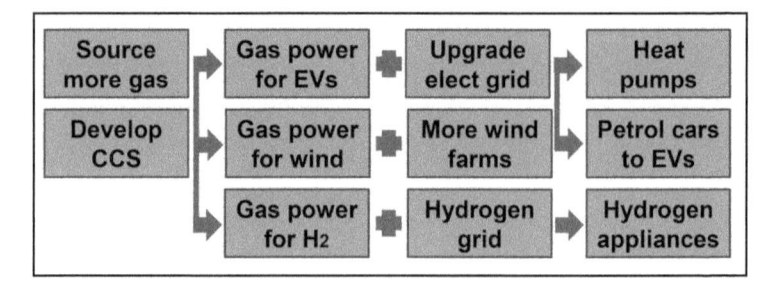

[Diagram:
Row 1: Source more gas → Gas power for EVs + Upgrade elect grid → Heat pumps
Row 2: Develop CCS → Gas power for wind + More wind farms → Petrol cars to EVs
Row 3: Gas power for H2 + Hydrogen grid → Hydrogen appliances]

Figure 29 Process diagram for implementing the key elements of the CCC strategy.

skills, logistics, funding streams and all the other aspects of infrastructure renewal present enormous problems, which would be compounded if other countries decided to follow a similar path on a similar timescale.

There are several activities on the critical path that must be advanced in series (see Figure 29). For example, there is no point introducing EVs in quantity until we have sourced the extra methane and built the power stations to generate the extra electricity, nor converting heating systems to use hydrogen until hydrogen mains have been installed. I reckon if we immediately put the UK on a command-and-control war footing for the next 30 years, achieving the CCC plan might be possible by 2050. But we tend to have a recession about every decade that demands a slowdown on investment, with the current COVID one probably lasting two or three years; and every large-scale public project inevitably takes twice as long and costs twice as much as originally envisaged. The whole process, prior to zero-carbon status, will generate an enormous amount of CO_2 emissions, which would take perhaps a decade to expunge in the fledgling zero-carbon world such that a net-zero UK would be unlikely before about 2080. There is no point setting the legal targets that the Climate Change Act does when no amount of government diktat can command the invention of new technologies nor find a magic wand to achieve a quick outcome.

And of course the public would be compelled to go along with the radical and expensive change to their lives. The CCC says that "people and businesses will choose to adopt low-carbon solutions, as high-carbon options are progressively phased out." We will "choose" new technologies, but only because Big Brother will take away all other choices. Think of what we would have to accept.

- An increase in carbon tax from £18/tCO$_2$ to £75/tCO$_2$ to deter us from using fossil fuels which will further undermine our economic base and exacerbate fuel poverty.
- The introduction of a Green Gas Levy (recommended from as early as 2021) which will also increase natural gas prices.
- We will be forced to buy EVs at £10,000 more than petrol ones, or do without.
- We will be forced to chuck out perfectly good gas heating and cookers and buy hydrogen ones.
- The alternative heating system is heat pumps which will cost around £15,000 to install.
- The streets will be dug up for at least a decade to install a hydrogen grid, improve the electricity grid for the extra EV-charging, electric heating and heat pump demand, and the decommissioning of the natural gas grid.
- Air travel will be priced out of the pocket of many families and businesses.
- Farmers will not be able to use cheap red diesel, will be forced to halve the number of methane-belching/farting cows (milk, butter, cheese and meat will increase in price accordingly) and a third of their land will be (effectively) requisitioned for biofuels and forestry. Laboratory grown meat will replace real meat.
- Local authorities will have to install expensive CCS facilities if they wish to build or extend waste treatment facilities, a cost that will be passed on to residents.

Renewables do not significantly reduce our reliance on fossil fuels, but by opposing fracking for gas and oil, and not mining our plentiful reserves of coal for steel and glass production, we merely increase our imports, with unnecessary increased costs as well as the increased emissions of transporting them. It seems that the CO_2VID virus affects the common sense in politicians and their academic advisors.

When energy costs began to rise, the UK government blamed the big six energy suppliers for not being competitive enough (it was nasty corporate greed that was the problem, not irrational energy policy) so it encouraged new players to enter the marketplace. But the prices continued to rise so it introduced cost caps to 'protect the consumer'. It is inevitable that too much government interference in market forces causes more problems than it solves, and so 25 energy suppliers went bust in 2021, with the resulting billions of pounds absorbed by the government at a time when it had already accumulated huge costs due to COVID-19.

To eco-moralists, it does not matter that we have no technologies to significantly reduce our carbon (dioxide) emissions, never mind achieve the utopian 'net-zero'. In fact, it is essential that we do not so that they can continue to preach that humans are inveterate fossil fuel sinners, without the moral compass to break the habit. Sitting on their moral high ground they are the ultimate hypocrites, their modern lifestyle being dependent on coal, oil and gas.

We will explore how ideological dogma trumps rationality in the next chapter.

CHAPTER 11
THE GARDEN OF EDEN COMPLEX AND 'WE'RE ALL DOOMED'
PSYCHOLOGY LESSON

For several years I tried to engage many intelligent people in rational discussions on climate change, but was frustrated by their lack of reason and their intolerance of alternative ideas. Political and deeply-held ideological world views seem to deter objectivity and reasonableness. This needs some explanation and this chapter considers why people hold their views on this subject so religiously and passionately, calling out those like me with a healthy and proper scientific scepticism as 'deniers of the science'.

The term 'denier' is usually an attempt to equate the person involved with a morally bereft Nazi Holocaust denier which is not only disingenuous but highly insulting to a Jew like Steve Koonin, who lost more than 200 of his relatives to the Hitler regime. Koonin was an advisor to the Obama administration and in late 2013 was asked by the American Physical Society, which represents 50,000 physicists, to review its 2007 statement on climate change. This was controversial because it contained the unscientific word 'incontrovertible'. Instead of just replicating the message of an IPCC Summary for Policymakers, as similar professional institutions had done, he assembled a panel of three IPCC editors and three sceptics to review the actual evidence cited in assessment reports. He concluded that the advice he had been giving the president of the USA on climate change was highly suspect because the evidence was far from clear. He set out his conclusions in a 2021 book *Unsettled: What Climate*

Science Tells Us, What It Doesn't, and Why It Matters, but by telling it as he saw it, he was branded by some as a denier.

This chapter explores issues such as the contention that there was some benign state of nature, or climate, or of humanity, in the past that was ideal but was despoiled by our present indiscretions/greed/selfishness, and that could be redeemed if we just stopped being indiscreet/greedy/selfish. In other words we could stop being sinful and re-enter the Garden of Eden, a sort of heaven on Earth. This seems to me more about religion than science; about utopianism than pragmatism. And why are some people so pessimistic about the future, the same question that Thomas Macauley asked himself in 1830: "On what principle is it that when we see nothing but improvement behind us, we are to expect nothing but deterioration before us?"

This is a long chapter because it is key to understanding why so many people get so passionate and, I would contend, irrational about climate change. Please excuse my extensive rant, but I believe it is extremely important.

Before the mid-20th century, the natural world was a dangerous place to be. Cities may have had more risk of pollution and epidemics, but we flocked to them because they were safer than the countryside with its wild animals, mosquito-borne diseases and greater exposure to a harsh climate. They offered hope. With new technologies a few intrepid adventurers explored 'Darkest Africa' and 'the Wild West' in the early 19th century, and the polar regions at the beginning of the 20th century. By the second half of the 20th century there was mass travel to all parts of the world, including wilderness areas, and we could 'get back to nature', just so long as we could retreat back to the safety of our high-tech modern world. Instead of fearing and fighting nature we could 'commune' with it. The word 'ecology' was invented in 1866 by Ernst Haeckel and a new philosophical and ethical tradition was born, that many people have taken to a religious level.

The three Abrahamic religions, Judaism, Christianity and Islam, were all products of hot countries. A lush garden was the idea of paradise, always wished for but rarely experienced. Adam and Eve, the first humans in this religious tradition, lived in the Garden of Eden, a natural state where they were naked, because lust and sin did not yet exist. God told them that they should not eat from "the tree of knowledge of good and evil", but Eve was tempted by a serpent to pick fruit from the tree and give it to Adam. Original sin was born, and because such mythology was made up by men, Eve got the blame for tempting Adam. Obviously women have weaker willpower to be sinful in the first place, and men are defenseless in the face of women's charms. As punishment from God, all women thereafter suffered pain in childbirth. This was a prelude to men and women taking from the tree of knowledge and being too clever for their own good. If only they had done what they were told without question, they would not have got things wrong.

The book of Genesis stated that all people should "have dominion over the fish of the sea, and over the fowl of the air, and over the cattle, and over all the Earth, and over every creeping thing that creepeth upon the Earth", basically giving them freedom to do what they wished with the rest of the natural world. But by the second half of the 20th century, with 3 billion people on the planet, it was not just the perfect garden that was being defiled by humankind, but the whole of creation, according to the new environmental movement. 'Pristine' nature could then be romanticised, and we could talk of it being in balance before nasty, greedy, sinful humans disturbed it.

A documentary series like the 2021 Attenborough/BBC *Perfect Planet* preached this message and sought to provide justification for radical measures to address our failings. For me, the best nature programmes show how dangerous the natural world is and how creatures struggle to survive, with extinction the default state, a fate which has befallen 99.99% of all species which have ever evolved. Nature is far more cruel than humanity could ever be. When David Attenborough was still a natural scientist, before he became a propogandist for green causes, he wrote *The Trials of Life* in 1990 which described species' struggle to pass on genes, following his wonderful *Life on Earth* 1979 and *The Living Planet* 1984. He joined the BBC the year I was born and it is sad to think that he caught the green malaise in later life. His boundless enthusiasm has been misdirected.

ENLIGHTENMENT SCIENCE VERSUS WISHFUL THINKING

I could write several books on the ideas of Enlightenment philosophers David Hume, who wrote *A Treatise of Human Nature* (1739–40), Adam Smith, *The Theory of Moral Sentiments* (1759) and *An Inquiry into the Nature and Causes of the Wealth of Nations* (1776), and the game-changing work of Charles Darwin, *On the Origin of Species* (1859). I would compare these with the ideologies of Karl Marx, *The Communist Manifesto* (1848) and *Das Kapital* (1867–1883), Friedrich Nietzsche, *The Birth of Tragedy* (1872) and *Untimely Meditations* (essays 1873–76), and Sigmund Freud, *The Interpretation of Dreams* (1899) and *The Psychopathology of Everyday Life* (1901). However, if I were to summarise the difference between the first three British philosophers and the three Germanic ones it would be that the former studied the actuality of human nature, economic mechanisms and the natural environment, and built on that foundation, while the latter despaired about the way the world was and sought radical new orders in a utopian fashion.

The former led to the success of democracy (individualism), free-market capitalism (consumerism) and ecology (humans as a part of nature), in contrast to the dire consequences of communism, national socialism (Nazi) and the pseudo-science of psychoanalysis. The former applied scientific principles while the latter peddled wishful thinking. It is my contention that there remains unfortunate remnants of utopianism among those who defend the climate change ideology with an unreasonable passion, sometimes supported by neo-Marxist ideology that contradicts Enlightenment values.

In classical Greece there were two competing philosophical traditions. Hedonism was good in one but bad in the other. The former was proposed by the Greek philosopher Epicurus (341–270BCE), from whom we get the modern word 'Epicurean', someone who enjoys good food. The Greek word *hedone* means 'pleasure', and this approach said that we should enjoy our lives, contentment coming from freedom from discomfort and pain. The Roman Titus Lucretius (c99–55BCE) further developed this in his work *De Rerum Natura*, or *On the Nature of Things*.

The alternative to this was the contention that the material world is inevitably corrupt and imperfect, while only the spiritual realm is perfect, so we must endure the pain of our sinful lives and hope for an eternity of peace after death in heaven. An emperor like Constantine found such a philosophy useful, since a pre-industrial economy could not provide wealth for everyone, just an elite top 10% or so. Christianity made it a virtue to be poor because, if one behaved oneself during one's life, despite a miserable existence, one could gain the reward after death. Much of the time this promise deterred peasants from revolting, so Constantine made Christianity the official religion of the Roman Empire in the 4th century. Islam, a word which means 'submission', provided a similar dictatorial discipline for Muslims, introduced by Caliph Abd al-Malik in the late 7th century (Tom Holland *In the Shadow of the Sword*). But we now know that there is nothing virtuous in being poor. As we saw in Chapter 7, extreme poverty should be eradicated throughout the world as a matter of urgency, and we have the resources and knowledge to do so.

Christian Rome destroyed any books it found which contradicted the official ideology, but a copy of *De Rerum Natura* survived and was discovered in 1417 in a German monastery. It influenced thinkers during the Reformation and the Enlightenment. Lucretius was a materialist who said that the world was made up of small particles combined in different forms, particles that today we call molecules. He said that the universe operates through physical principles, guided by *fortuna* (chance), and not the divine intervention of the gods. He wrote: "If you possess a firm grasp of these tenets, you will see that Nature, rid of harsh taskmasters, all at once is free, and everything she does, does on her own, so that gods play no part... O miserable minds of men! O hearts that cannot see! Beset by such great dangers and in such obscurity you spend your lot of life! Don't you know it's plain that all your nature yelps for is a body free from pain and, to enjoy pleasure, a mind removed from fear and care?... The elements of things do not collect and order their formations by their cunning intellect, nor are their motions something they agree upon or propose; but being myriad and multi-mingled, plagued by blows and buffeted through the universe for all time past, by trying every motion and combination, they at last fell into the present form in which the universe appears."

This is a basic description of the evolutionary process, 1,800 years before Darwin. Armed with this newly rediscovered philosophy, science in the Industrial Revolution presented the promise that everyone could share in wealth, even if it took many years for the benefits to work their way through all sections of society.

Allied to this was the development of democracy, giving power to the people, not as an angry mob or with just the ability to post votes in ballot boxes, but as individual traders and entrepreneurs responding to the demands of consumers to improve their lives in a way that was not over-controlled by a divinely appointed monarch or a bureaucratic state. A bottom-up free market system was much more sensitive to people's needs than a centralised dictatorship. Capitalism fuelled this, by allowing individuals and companies to borrow money to advance their ideas quickly, using previously accumulated capital (property and other assets) as a guarantee to lenders in the event of failure. Ordinary people could invest their savings in the stock market, helping to develop new commerce for the common good, and insuring themselves for a rainy day or old age using the returns on investments. The state could cream off a bit of the profit in taxes to provide social services like drinking water, metalled roads, sewerage systems, street lights, a police force and elementary education. As countries became richer, they could afford more social services until today the public sector can utilise around half of the wealth of a developed country.

The modern environmentalist imagines a default state of order and harmony in nature. When anything goes wrong, we have a natural inclination to apportion blame and declare people stupid or even bad. This saves us the time trying to understand the person or the issue. As Rosling said: "We like to believe that things happen because someone wanted them to, that individuals have power and agency: otherwise the world feels unpredictable, confusing and frightening." Unfortunately the world is all those things. Pinker: "A major breakthrough of the Scientific Revolution – perhaps its biggest breakthrough – was to refute the intuition that the universe is saturated with purpose… While few people believe that accidents or diseases have perpetrators, discussions of poverty consist mostly of arguments about whom to blame."

Evolution has no goal and no plan. Organisms which adapt and work in practice survive and reproduce, and it is the same with human endeavours and ideas. This lack of control and certainty is scary and it is no wonder that many people still cling to religions where a saviour with superpowers can be called on, even if a deity works in mysterious ways – the calls for help seem to go unanswered. Many people appear compelled to believe that humans can control the climate, or the economy, or any other facet of existence, or predict the future, while reality suggests that the best we can do is respond to situations as they occur on the basis of past experience, which includes maintaining reserves to help us cope. Being resourceful, alert and open-minded is the best we can do. Utopian ideals by themselves are useless, because pragmatism must be applied to aspiration.

ROMANTICISM AND CONTRA-ENLIGHTENMENT

Simon Sharma, in his well-researched and thoughtful 2020 BBC documentary *The Romantics and Us*, contended that the modern world was born at the end of the 18th century with the emergence of Romantic poets, artists and musicians. "A generation of artists working 200 years ago around the time of the French Revolution… [with a] new religion of insurrection and agitation, in which everyone could take to the street to fight for freedom, equality and justice… We still think with their mind, we feel with their emotional heartbeat, we see with their eyes, and we listen with their ears." Well, Mr Sharma, you might, and the XR protestors you use to illustrate your point certainly do, but we are not all old Romantics like you. Please count me out of the "us" in the title of your programme.

Sharma claimed that his subjects were "the artists who created the secular icons of our democratic world", but the Romantics, like William Blake, "targeted what he felt was the true enemy of his age, the Enlightenment, whose faith in pure reason and science cut man off from his sacred potential". In other words, the Romantics valued romance and emotion over science and reason, while I contend that the modern, civilised, democratic world needed the latter to prevail. Liberty, fraternity and equality were not the products of the French Revolution, but anarchy. David Hume would have countered by recognising that, while we are creatures of passion and emotion, reason must shape the things we do to obtain an emotionally satisfactory outcome.

Despite his comfortable lifestyle as a scholar and broadcaster, Sharma has a poor view of the modern world. "This idea, forged at the beginning of our modern understanding of the human mind has become one of Romanticism's greatest legacies, and it feels as relevant now in our helplessly tumultuous times as it ever did… As we become ever more materialistic at the expense of our own happiness and the health of the natural world, questions remain as to whether it is something we have come to realise too late. We are living in an epoch of self-obsession."

The jettisoning of Enlightenment for Romantic ideals is not a legacy I value, nor would advocate. Passionate demonstrations, 'righteous indignation', celebration of high-brow arts, soul-searching and endless despair about the problems facing us will not solve anything. Scientific principles and industrious endeavour hold the key. Romanticists are not pragmatists and are slaves to their pessimism. They should ask poor people what they think about material comforts before they condemn us all as "hopeless materialists". Great hope comes from being resourceful.

William Wordsworth fled from France in 1793, disillusioned by the chaos after the revolution, and took solace in the British countryside, inventing a romantic notion of once-threatening landscapes, where they became a mentally relaxing foil against "the brutal materialism of the modern world". It is fitting that Sharma recruited Attenborough to read one of Wordsworth's poems.

Many of the Romantics were drug addicts and some ended up in mental asylums, such was their despair ("their innate sense that madness was actually an essential element of creative genius"). The romantic heroes are, for the most part, tragic characters, one might say society's losers, but they are elevated by the modern intellectual's nostalgic view of them. Romanticism relies on reinventing the past as something better than it was, and imagining possible futures, often with fear of their consequences. Environmentalism fits in with this, where a Garden of Eden natural perfection and a stable pre-industrial climate are the basis on which to imagine computer-modelled cataclysmic futures. It is not science; it is mythology.

Sharma noted that 'nostalgia' was originally defined as a serious, but treatable, medical condition. He traced the emergence of it in nations which changed political and cultural status, and whose outlet was the poetry and music of his Romantic heroes: Scotland after the union with England in 1707 (Robert Burns), the German provinces in the process of unification up to 1871 (Jacob and Wilhelm Grimm) and the centuries-long struggles of Poland to resist the domination of Germany (formerly Prussia and Austria), Russia and Napoleon's France (Frederyk Chopin).

The original 1812 folk story collection of the brothers Grimm was truly grim. The savagery, mutilation, incest and cannibalism so reflected the gruesome reality of the pre-industrial Little Ice Age world, that today we can only cope with sanitised Disney versions. The sordid possibilities of a manufactured, nostalgic, national culture were obvious in 1930s Nazi Germany, and Sharma's joy at the British patriotic pomp and circumstance at the Last Night of the Proms concert is at odds with the currently fashionable spotlight on the downside of imperialism and empire.

In Scotland he chose to show people crying into their beers

listening to a Burn's song which imagines the prelude to the Battle of Bannockburn in 1314, one of the very few battles that little Scotland won against its much larger neighbour. Today this nostalgia fuels nationalism, which contains a (minority?) anti-English racist element, especially towards 'posh' English people. In reality, 'independence' would not be 'freedom', but merely the transfer of effective sovereignty from London to Brussels, swapping the pound for the euro. Ironically, the economic strength to effect the very expensive separation of Scotland from England might come from its fossil fuel reserves if green ideology did not pervade Scottish nationalism, which bans fracking. The wealth of the nation has always lain in fossil fuels, from James 'paraffin' Young and shale oil, to coal mining and North Sea gas and oil.

West-central Scotland, along with the island of Ireland, is also cursed by the romantic celebration of the Battle of the Boyne in 1690 by Orangemen, setting modern Protestants against Catholics, with the euphemistically named 'Troubles' when, between 1969 and 1998, around 3,500 people were murdered in sectarian violence (the hatred and violence have still not completely gone away). I suppose this is slightly less absurd than the Shia/Sunni divide, based on the mythological accounts of the Battle of Karbala in 680. The facts have inevitably been distorted by the romantic literature of the following centuries. Muslim on Muslim violence is much more prevalent than radical Islamic terrorism outside the Muslim world.

Scottish nationalists might bemoan their alleged mistreatment by their southern neighbours, but this is nothing compared to the tribulations that Poland has suffered over the last two centuries, with millions of its people murdered by aggressors. Today Poland is under pressure from the Franco-German dominated EU to abandon its cheap domestic coal for inefficient and expensive renewables, while Russia seeks to supply it with essential oil and gas. To the wary Pole, the old enemies in different clothes are at the door and with new threats. The country's economic self-sufficiency is compromised by green ideology, with consequent pressure on democratic institutions.

Just as the first industrialised nation was the first major empire to be able to abolish slavery, it was also the first to formulate mass participation sports – football/soccer, rugby, golf, cricket, tennis – initially among the emerging middle classes with spare time for leisure activities, and then manual workers. As recently as my youth, men would participate in the drudgery of commercial necessity for five-and-a-half days, be passionate on the football terraces on a Saturday afternoon (I remember the thrill of being among the 136,505 people who watched Glasgow Celtic versus Leeds United at Hampden Park in 1970) and be devout about the passion of religion on a Sunday, when few secular activities were permitted and I attended four religious activities. It is safer to restrict natural passions to national/tribal sports, which are governed by clear rules of combat and in the end don't really matter, to avoid harmful nationalism/racism/bigotry and real conflict.

The ability for the secular/scientific to prosper alongside the spiritual/mythical was a key to democratic institutions emerging in Christian countries, especially within the reformed religion with its so-called 'Protestant work ethic'. This is more difficult for Muslim nations, where religious rituals take place throughout the day, every day. They must be passionate about religious mythology all the time, and the constant exposure to dogma leads to a more fatalistic world view. Indonesia, Tunisia and Malaysia are the only majority-Muslim democracies, at the western and eastern extremes of Islam, far from Medina.

Discussion: Human beings possess both rationality and emotion. We can analyse a problem and review possible solutions ad nauseam, but it requires an emotional spark to initiate action. But what can happen if we respond emotionally without thinking first? Is it more difficult to be rational or emotional? Can our most valued experiences be the result of random emotions, like 'falling' in love? Is serendipity – taking advantage of chance finds – more enjoyable than seeing a carefully engineered plan come into place?

CLIMATE CHANGE, NEO-MARXISM AND THE US GREEN NEW DEAL

Winston Churchill once said that democracy is the worst form of government, apart from all the rest, and the same could be said of capitalism in terms of economic systems. Capitalism has endured, not because it is perfect, but because it is the best we have found so far. When it stutters and financial crises arise, instead of contributing to the constant challenge of reform and improvement, some have sought more radical solutions. The horrors of the Nazis are still fresh in Western human memory, so few people today promote the right wing ideology that was developed from Nietzsche's ideas (even if some of them were possibly valid). The horrors of Stalin's Soviet Union or Mao's China were just as bad, but there are many Western intellectuals who claim that these were poor interpretations of Marxist ideology and that left wing neo-Marxism is the answer. They insist that this time they will get the socialist solution right.

A utopian, neo-Marxist agenda has long been present in Western academia. I was an enthusiastic socialist at university in the early 1970s, with a poster of Che Guevara on my wall, until I entered the commercial world. The agenda emerged in US Democratic Party strategy as an alternative to Republicanism, with which Donald Trump gained power as an alternative to the previously unresponsive Washington establishment of the Obama era that seemed to have paused the economic progress of middle America. The champion of the Green New Deal is young Congresswoman Alexandria Ocasio-Cortez, who has been in politics for only a few years. For seven years before that she was a waitress in New York and was a Bernie Sanders supporter. She is allied to an organisation called New Consensus which drafted the US Green New Deal resolution. Climate change is seen as proof of humankind's failings. Donald Trump might make a comeback as the consequences of the Democratic 'green' energy policies bite in America.

As she said, "Our greatest existential threat is climate change. And so to get us out of this situation, to revamp our economy, to create dignified jobs for working Americans, to guarantee healthcare and elevate educational opportunities and attainment, we will have to mobilise our entire economy around saving ourselves and taking care of the planet." The change demanded is radical and revolutionary. Climate change is the spark to light the blue touch paper.

There are five basic goals of the Green New Deal:

- Net-zero emissions
- Good, high wage jobs
- Transformed infrastructure and industry
- A clean and sustainable environment
- Justice and equity

AO-C, as she is known, is all about aspiration, not details. "A lot of what the Green New Deal is about is shifting our political, economic and social paradigms on every issue because we don't have time to wait, we don't have time for five years for a half-baked, watered-down compromise position… To think that we have time is such a privileged and removed-from-reality attitude… Climate change is different because we have an expiration date. The IPCC report says we've got 12 years to turn it around. We are going to be the frog in the pot of boiling water." Of course the IPCC does not say this. It took 2030 as a convenient date and relatively foreseeable timescale from *IPCC SR1.5* in 2018 to consider various scenarios, but it does not say that the world will end then if we do not take immediate action. As we saw in Chapter 1, activists take IPCC information and bend it to justify their political views.

She emphasises that it is not just about addressing environmental problems. "Here's the deal. We could solve all of the environmental issues in the world. If those climate policies and solutions are drafted on to the existing framework of economic injustice then we will perpetuate our social problems." Desmond Drummer of New Consensus: "It's about reshaping the entire economy; a fossil fuel economy that is designed to exploit and extract requires disposable people and disposable places. The Green New Deal says no more disposable people." Drummer wrongly contends that someone "designed" an exploitative economy with mal-intent (a Marxist idea), but economies, like everything else in nature and human society, evolved by a process of trial and error such that the most practical things survived, in a flawed but workable system. There is no 'malevolent' controlling hand, and there should be no 'benevolent' one if it has the sort of all-powerful controlling hand that would be needed to implement the Green New Deal.

A country like the Soviet Union, which tried to plan its economy on Marxist principles, was most certainly malevolent. In the 1930s intellectuals came to the countryside from Moscow with the message that property is theft and success is oppression, and sent the best farmers to Siberia because of their crime of working hard to accumulate capital in the form of property and livestock. Their parents had been serfs, essentially slaves, to the Russian aristocracy, and its removal had given them entrepreneurial opportunities. But now the inequality caused by their success was deemed unfair under Marxism, and the loss of their skills to the economic system led to the starvation of 6 million people, who were certainly then all equal. Mugabe's Zimbabwe, formerly the 'breadbasket of Africa', suffered the same fate when it expelled white farmers. Its economy went into free fall and inflation mushroomed. Equality of opportunity is a noble cause, but equality of outcome is only experienced when social disruption drops everyone to the same basic level.

'Nature' will inevitably throw up individual people with a variety of attributes on a spectrum of cleverness, physical and mental strength, beauty, personality types, etc., and then pure luck will throw a few hard balls. There is a limit to the extent that 'nurture' can improve a perceived weakness in any one of these. This is not fair, but fairness does not feature in nature and the evolutionary process. Social constructionists who look for victims will inevitably also seek perpetrators, but the blame game is a fruitless fool's venture, and very divisive. We should accept what we see before us, and work together to make the best of it.

AO-C wants to subvert the market, and presumably take control of it: "The one thing that we cannot rebuke, and the one thing that we cannot deny, is that climate change is a problem of market failure and externalities in our economics. Moreover, Exxon Mobil knew that climate change was real and man-made starting as far back as 1970. The entire United States government knew that climate change was real and human-caused in 1989, the year I was born. The initial response was 'let the market deal with it'. Forty years and free market solutions have not changed our position." Market economics, and the multinational companies which support it, are defined as evil powers because all power corrupts in a Marxist world view.

Once again, her starting point is rhetoric and not facts. The nasty oil companies knew they were doing wrong as far back as 1970? But the false alarm then was global cooling. Evil politicians knew the truth in 1989? But the first IPCC assessment report in 1990 concluded that it did not have enough evidence to connect human emissions to climate change. It was *IPCC AR2* in 1995 that tentatively said: "the balance of evidence suggests a discernible human influence on global climate". It was not until *IPCC AR4* in 2007 that most of the warming since 1950 was attributed to human emissions. She uses the 'deny' word to convey that she will not accept any hint of criticism of her hypothesis.

AO-C is not old enough to have made her own mistakes and learn from them, and clearly has not studied enough history to learn from other people's mistakes. She is making up her own version of history to justify her ideology, just as activists make up their own version of climate science to 'prove' that things are so bad that we cannot lose by adopting radical social change.

Rhiana Gunn-Wright of New Consensus has the strategy: "When we have the full workforce at the ready, what we can do is unknown." The first step is to provide healthcare, childcare and training to those who are disadvantaged. They will then have the security and skills to fill new jobs in renewable and sustainable industries. But where will the money and resources come from to provide enhanced employment rights and benefits, retrain a generation (at a moment's notice) and subsidise inefficient and expensive renewables? The resolution envisages, "enacting and enforcing trade rules, procurement standards, and border adjustments with strong labor and environmental protections to stop the transfer of jobs and pollution overseas; and to grow domestic manufacturing in the United States". But the transfer of jobs overseas has been largely caused by adopting expensive renewables and making domestic energy costs uncompetitive, such that manufacturing becomes uneconomic. This strategy would only make the situation worse.

AO-C seeks popular appeal for the revolution from those who she believes are discriminated against by the system and are dispossessed. The resolution itself makes clear who her army will be: "indigenous peoples, communities of color, migrant communities, deindustrialised communities, depopulated rural communities, the poor, low-income workers, women, the elderly, the unhoused, people with disabilities, and youth (referred to in this preamble as 'frontline and vulnerable communities')". That leaves white, middle-aged, middle class, able-bodied men, living in cities, as the enemy. Surely this is a racist, ageist, classist, anti-ability, misandrist and anti-urban sentiment? She talks about "the poor", but Matt Ridley would remind her that "The [American] middle class of 1955, luxuriating in their cars, comforts and gadgets, would today be described as 'below the poverty line'... Today, of Americans officially designated as 'poor', 99% have electricity, running water, flush toilets and a refrigerator; 95% have a television, 88% a telephone, 71% a car and 70% air conditioning."

Despite living conditions in a Western country like America being better on average than ever before, there are still many issues to be addressed. We should strive to improve the lives of many unfortunate, disadvantaged people, but revolutionary change in the past has had dire consequences. Successful transitions happen by stealth, not by impulsive and impatient leaps of faith, because some initiatives fail and some are successful, and it is difficult to predict which will be which. I learned in business that life is about trial and error, learning from our mistakes as we go. She who never made a mistake never achieved anything, but America and the Western world cannot afford to be the guinea pig for AO-C to repeat the socialist mistakes that have failed elsewhere, and which were seen most recently in Venezuela, where a country rich in natural resources was bankrupted by ideology.

The UK equivalent is the Environmental Justice Commission, part of The Progressive Policy Think Tank, which published its

cartoonsbyjosh.com

plan on 14 July 2021, *Fairness and opportunity: A people-powered plan for green transition*. Its co-chairs are Green Party MP Caroline Lucas, Labour MP Hilary Benn and ex-Conservative Party MP Laura Sandys, the last of these, according to *The Guardian*, seeing the light in April 2019 when she called for a Green New Deal in the UK.

> **Discussion: Consider the following statements. Identity politics involves placing people into categories in society by sex, class, wealth, race, age, nationality or any other attribute, and branding them with perceived characteristics, good or bad. This is an affront to the humanist principle of the inherent value of each individual. The resultant tribalism is guaranteed to promote social division and hatred. But perhaps governments need to classify groups of people in order to direct aid to those who need it most.**

WORLD GOVERNMENT

The 1966 Beatles' song 'Taxman' was a protest about the high tax rate of 95% for rich people: "Should five per cent appear too small, be thankful I don't take it all". This was a time when many of the most talented people in British society, like scientists and doctors, joined the 'brain drain' to live in more liberal countries like the USA. Socialism in the UK led to economic breakdown in the 1970s which led to the neo-liberal reforms of the 1980s, with less state intervention in commercial activities. As a philosophy, socialism only flourished where people were contained within the jurisdiction that imposed it, as in the Soviet Union and its satellite states. The Berlin Wall fell after East Germans could see on television how much better the capitalist West Germany fared. For neo-Marxism to work it would have to be applied globally so that there could be no escape from it. In a world where some multinational companies have bigger economies than individual countries, many see the UN as the vehicle to impose a more 'just and equitable' world order. Climate change activists see the UN as the body to enact their policies, because there is little benefit in a few countries shrinking their emissions if most do not. They want a one-world-fits-all solution.

The UN was set up after the Second World War as a forum which it was hoped would keep world peace, but it quickly became clear that it was almost as impotent in this respect as the League of Nations that it replaced. The most powerful economies and political ideologies did what they wanted and, even today, a single veto by a member of the security council can stop any noble or sensible initiative. So the UN moved to its secondary goals such that by the 1970s its budget for social and economic development was much greater than its peacekeeping budget. Under its auspices are the International Court of Justice, the International Fund for Agricultural Development, the World Food Programme, the United Nations Educational, Scientific and Cultural Organisation (UNESCO), the World Meteorological Organisation and several more. The US-based World Bank and the International Monetary Fund are parallel organisations which were persuaded in 1989 to add environmental strategies to policies for international development.

Canadian multi-millionaire Maurice Strong, who died on 27 November 2015 (obituary in *The Telegraph* 6 December 2015) played a key role in placing environmental issues at the heart of the UN. He once said, "Isn't the only hope for the planet that the industrialised nations collapse? Isn't it our responsibility to bring that about?" (The quote is from Elaine Dewar's book *Cloak of Green*.) He believed that a new way of organising the people of planet Earth had to be found, following the 1930s recession that had demonstrated the problems with capitalism, and the pollution of the Industrial Age West that had shown our proclivity to defile the planet. In 1972 he chaired the first UN Conference on the Human Environment, which led to his appointment as the first director of the UN Environment Program. Along with the World Meteorological Organisation, UNEP set up the Intergovernmental Panel on Climate Change in 1988.

No politician wishes to be seen as environmentally unfriendly, so in 1992 Strong managed to persuade 179 nations to sign up to Agenda 21 at the Earth Summit in Rio de Janeiro. The first UN climate change conference (Conference of the Parties, or COP) took place in Berlin in 1995 but, like most UN ventures, national interests generally prevailed. Even the Paris COP in 2015, which was hailed as a great success, merely produced a resolution that effectively declared that each country would do the best it could to reduce emissions, if it could be bothered. Critics declare that 30,000 people fly in to international venues, achieve very little, and preach to the rest of us that we should fly less to reduce emissions.

Developing nations claim that Western countries caused the problem of emissions and climate change in the first place, and that they should pay the high cost of reorganising developing economies to use renewables rather than fossil fuels. No one wants to suffer the inevitable economic consequences associated with the radical change that is prescribed. The UN is doing great work in international aid and development, but world government is a long way off. This is great relief to those who believe that democracy needs manageable, discreet units in which to operate effectively, like nation states, and that world government would inevitably be too insensitive and dictatorial.

Incidentally, Strong later moved to China after being implicated in the Saddam Hussein 'oil for food' scandal with respect to an allegedly illicit $1 million payment to him, and advised the government there on climate change and carbon trading. His relative, Anna Louise Strong, was a Marxist who had spent two years in China with Mao Zedong and Zhou Enlai during the Cultural Revolution. This was an example of a rich man using his wealth to right the wrongs of the world as he saw them, but losing focus on wealth generation so that he had to turn to 'dodgy' dealings to continue funding his activities.

Discussion: It was once said that socialism starts with compassion and ends with compulsion, and that a leader who is guilty of evil deeds doesn't get up in the morning aiming to promote evil, but to make the world

a better place. Is it inevitable that too much power at the top of society leads to a cruel and insensitive dictatorship, no matter the best of intentions? What would you want to do if you ruled the world? Would you want to rule the world?

OCCUPYING THE MORAL HIGH GROUND INEVITABLY LEADS TO HYPOCRISY

In Chapter 1 we came across one of Greta Thunberg's advisors, Kevin Anderson, Professor of Energy and Climate Change at the University of Manchester, and at the Centre of Sustainability and the Environment (CEMUS) at Uppsala University in Sweden. He describes himself as a 'progressive lefty' and he places the blame for emissions production firmly at the feet of the rich: "The taboo issue of the huge asymmetric distribution of wealth underpins the international community's failure to seriously tackle climate change." He says we need to "disturb the dominant socio-economic paradigm of ongoing growth, with resources, power and CO_2 skewed to a privileged few".

He highlights the hypocrisy of this "few", noting the "1,000 to 1,500 private jets flying the morally bereft to Davos to discuss issues around climate change". He is disparaging about what he calls the climate glitterati: "Michael Bloomberg, Leo DiCaprio, Nick Stein, Christiano Figueres, Al Gore and Mark Carney, with huge carbon footprints supported by a cadre of senior climate academics, and it's about time we called these people out". He claims that COP conferences have the same hypocrisy, where people talk about speculative and largely irrelevant issues like negative emissions, carbon offsetting, geo-engineering, carbon capture and storage, electric planes and green growth 'within the system', when a more fundamental reorganisation of society is essential.

He agrees with part of my analysis in Chapter 10 that much climate change policy is fruitless greenwash. Offsetting is just paying a poor person to diet for us. Emissions trading involves so many permits that the price of carbon stays low. We "plant a tree so we can expand an airport. Geo-engineering is a sticking plaster on gangrene". Negative emissions technologies have not yet been invented but the problem is being passed on to future generations "because we can't be bothered doing anything else". He notes that renewables have limited utility: "There are big issues around storage and intermittency to some extent, and frequency control, and all those things. They're all resolvable in a reasonable time frame." But his time frame is immediate, and speculative technologies take decades to prove and roll out. Anderson is living in an academic world, a theoretical world, a modelled world, and not the world I earned my living in where pragmatism rules.

"People like me [rich people?] love the idea of a carbon price because we can afford it. I can just pay the extra on the price of a flight or a fancy car. Carbon pricing means nothing to me." At least he recognises his own hypocrisy, and so far he has not adopted the measures he thinks necessary, presumably because he is waiting for the revolution to start first. "Thus far there has been a litany of technocratic fraud. We have not seriously tried to cut our CO_2."

Anderson was previously Director of the Tyndall Centre for Climate Change Research, an organisation founded in 2000 and funded by the Natural Environment Research Council (NERC), the Economic and Social Research Council and the Engineering and Physical Sciences Research Council, all UK government quangos with a combined annual budget approaching £2 billion, much of which goes to academic climate change research. Perhaps it only goes to researchers who accord with the UK climate change orthodoxy. He wrote in 2010 that, "there is now little chance of maintaining the rise in global mean surface temperature at below 2°C, despite high level statements to the contrary... 2010 represents a political tipping point". He reckoned that there needed to be a fairly immediate 9% annual decrease in emissions by the advanced economies to try to stabilise CO_2 in the atmosphere such that they would peak at 450 ppm, but noted that UK emissions had increased by 18% since 1990, a figure which included emissions embedded in imported goods. He estimated that, by 2020, there was a 10%–60% chance of the average world temperature exceeding 2°C above pre-industrial, and a 50% chance of 4°C by 2050. The first date has already passed, and he is bound to be 50% right in 2050.

His neo-Marxist approach is revealed in his command-and-control solutions. Rich people who can afford more expensive technology like efficient fridges and electric cars **should be made** to get rid of their old CO_2-profligate devices. They **must** immediately abandon their large houses, holiday homes, second homes, prestige cars, SUVs, multiple car ownership in families, highly mobile lives, frequent flight business and first class travel and their high levels of consumer goods. We need to "**manipulate the system** and actually, maybe, **be much more revolutionary**". Just now "we have a liberalised choice system so we can buy crap if we really want to". He of course must be the paragon of virtue who can define what "crap" we should not be allowed to purchase.

Michael Moore is another 'lefty' but who has dared to contradict the orthodoxy by showing in his 2020 film *Planet of the Humans* (available on YouTube) that renewables do not significantly reduce our reliance on fossil fuels, and that they are harmful to the environment. He claims that they are a right-wing scam to make money and that only radical political and economic change will solve the 'climate crisis'. The private sector is certainly not going to refuse government handouts and a guaranteed price for generating electricity, whether it is used by the public or not, depending on how much the wind blows and the sun shines. Big energy companies can hedge their bets, depending on public demand and government subsidies, happy to profit from both fossil fuels and renewables, knowing that world demand for fossil fuels is guaranteed to grow for many decades.

Solar energy may not be a right-wing scam, but it was first promoted by energy companies worried that oil and gas reserves might soon be exhausted, as Charles C Mann pointed out in his thoughtful analysis of the two contrasting approaches to perceived ecological issues, *The Wizard and the Prophet* 2018. "Exxon became, in 1973, the first commercial manufacturer of solar panels; the second, a year later, was a joint venture with the oil giant Mobil… By 1980 petroleum firms owned six of the ten biggest US solar firms, representing most of the world's photovoltaic manufacturing capacity." He cited an article by British geologist Colin Campbell and the French petroleum engineer Jean Laherrère in *Scientific American* in 1998 stating that the global petroleum output would be in permanent decline by 2010, just as the hydraulic fracturing revolution was about to take off, proving them 100% wrong. It would be counterproductive for any energy company to challenge the established 'wisdom' on climate change, or admit that their expensive 'renewables' do not have any significant effect on a 'net-zero' energy policy. It makes sense to sit back and maximise the profits from our foolishness and hypocrisy.

Discussion: Is it possible that the climate change agenda is, at one and the same time, a left-wing construct by neo-Marxists to justify radical changes in society and a right-wing enterprise by the promoters of an eco-industrial complex? If it is, who will win and who will lose in such a situation? Have politicians lost control, assuming they were ever in control? Have they released a beast? Should we believe anyone who predicts the future, regarding themselves as a modern-day Nostradamus?

GREENISM: THE NEW QUASI-SCIENTIFIC, QUASI-RELIGIOUS IDEOLOGY

Patrick Moore was a co-founder of Greenpeace, and he claims the only one of its directors with a scientific background. After successful campaigns against nuclear bomb testing, whale hunting and baby seal culling, he left the organisation when, "The peace in Greenpeace had faded away. Only the green part seemed to matter. Humans, to use the Greenpeace language, had become the 'enemies of the Earth'. Putting an end to industrial growth and banning many useful technologies and chemicals became common themes of the movement. Science and logic no longer held sway. Sensationalism, misinformation and fear were what we used to promote our campaigns." (Should Friends of the Earth be renamed Enemies of Humanity? 'Misanthrope' is the hatred of the human race.)

"The final straw came when my fellow directors decided that we had to work to ban the chemical chlorine worldwide. They named chlorine 'the devil's element' as if it were evil. This was absurd. Adding chlorine to drinking water was one of the biggest advances in public health, and anyone with a basic knowledge of chemistry knew that many of our most effective pharmaceuticals had a chlorine component. Not only that, if this anti-chlorine campaign succeeded it wouldn't be our wealthy donors who would suffer. Wealthy individuals and countries always find a way round these follies. The ones who suffer are those in developing countries." Moore's books include *Confessions of a Greenpeace Dropout – The Making of a Sensible Environmentalist* 2010 and *Fake Invisible Catastrophes and Threats of Doom* 2021.

Humankind is back to where it was in the Bible: an inherently sinful bunch, heading to apocalypse. The truth is less dramatic but more complex. Our prosperity depends on sensible stewardship of the environment we live in. As we build up our resourcefulness, we have proved that we can do this better. And for some time to come fossil fuels will be our key resources. The renewables industry makes a limited reduction in emissions, wastes money and harms the environment. This is a faith we can do without. Radical, unscientific, evangelistic organisations like Greenpeace and Friends of the Earth should not be granted uncritical platforming by the media. They are poisoning the minds of impressionable young people.

ACADEMIA AND THE RUSH FROM LIBERALISM AND CONTROVERSY

Joanna Williams (*Academic Freedom in an Age of Conformity*) talks about 'woke conformity' that has long pervaded polite society. "Our universities have long since abandoned a commitment to intellectual freedom and the unhindered pursuit of truth. Our schools and museums swapped the passing on of knowledge for moralising more than a generation ago. Theatres and art galleries were converted to identity-driven, rather than artistic, goals long before Gen Z was even born... In a letter in *Harper's* magazine, signatories including Noam Chomsky, Malcolm Glaswell and Gloria Steinem decry the 'vogue for public shaming and ostracism' and make anew the case for free expression. In response, the letter has been roundly criticised as a temper tantrum by the privileged who find themselves, for the first time, subject to criticism from a younger, less powerful, more radical, more social-media savvy generation."

A study by Exeter, Keele and Aston Universities of Extinction Rebellion protesters in London in 2019 and 2020 found that about 85% had a university degree, three-quarters were from the prosperous south-east of England and two-thirds were women. The people who benefit most from the privileged lifestyle provided by liberal, capitalist democracies seem to be its biggest detractors. Arts courses, where neo-Marxist ideas flourish, are largely filled with women. Jordan Peterson reckons that young men are deterred by the (bizarrely flawed in his opinion) contention that most societal problems are the result of a malevolent patriarchy. They are made to feel guilty for being men. Women are the victims. Much of the media is populated by arts graduates, some of whom seem to lack the common sense and practical application of scientific rigour, and are attracted to more nebulous, aspirational ideologies. Journalists love climate change, because they can claim the authority of science to wallow in sensationalism and speculation. They do not have the ability to check out the scientific/academic studies they cite, even if they wished to.

And, of course, practically all climate 'scientists' operate in an academic environment. I have listened to a number of researchers talk about climate change at Meetup groups. They are candid about the uncertainties in their own discipline – there must be uncertainties or they would not obtain research funding to explore them – but are uncompromisingly loyal to the belief that humans are causing dangerous climate change. I find that I am more knowledgeable on other aspects of the subject and point out clear flaws, discrepancies and inconsistencies that question their faith, to which they have no answers. Often they ask me to contact them after the event with my queries and they will get back to me. But they never do, because they cannot face the rigours of real scientific debate, and probably cannot raise possible fundamental flaws in climate science orthodoxy with their colleagues for fear of scorn. They retreat to their safe, cloistered complacency. Science only thrives with controversy and disagreement, however uncomfortable this is, and is choked within a PC straitjacket.

Stephen Pinker reckons that many intellectuals suffer from 'Progressophobia'. "Intellectuals hate progress… It's not that they hate the fruits of progress… It's the idea of progress that rankles the chattering class – the Enlightenment belief that by understanding the world we can improve the human condition… In *The Idea of Decline in Western History*, Arthur Herman shows that prophets of doom are the all-stars of the liberal arts curriculum, including Nietzsche, Arthur Schopenhauer, Martin Heidegger, Theodor Adorno, Walter Benjamin, Herbert Marcuse, Jean-Paul Sartre, Franz Fanon, Michel Foucault, Edward Said, Cornel West, and a chorus of eco-pessimists.

"The left tends to be sympathetic to yet another movement that subordinates human interests to a transcendent entity, the ecosystem. The romantic Green movement sees the human capture of energy not as a way of resisting entropy and enhancing human flourishing, but as a heinous crime against nature, which will exact a dreadful justice in the form of resource wars, poisoned air and water, and civilisation-ending climate change. Our only salvation is to repent, repudiate technology and economic growth, and revert to a simpler and more natural way of life... Reason, science and humanism – these ideals are treated by today's intellectuals with indifference, scepticism, and sometimes contempt."

It seems that the privileged literati are undermining the pillars of the enlightened, liberal, capitalist society that pays for their privilege. With the crisis of confidence in their own social tradition, they are grasping failure from the jaws of success. The main beneficiaries of this contra-Enlightenment are the despots in authoritarian regimes in places like Russia, China and Saudi Arabia, who prosper from our foolishness.

CHAPTER 12
JOLLY HOCKEY-STICKS AND CYCLES WITHIN CYCLES
POLITICS, MATHEMATICAL GRAPHS AND HISTORY LESSONS

ARE TEMPERATURE AND CARBON DIOXIDE LEVELS UNPRECEDENTED?

Every organisation must meet its targets. Having been in business for ten years, the UN IPCC had not come up with the goods: proving conclusively that human emissions cause global warming. Its first report, *IPCC AR1* in 1990, concluded: "There are many uncertainties in our predictions, particularly with regard to the timing, magnitude and regional patterns of climate change, due to our incomplete understanding of sources and sinks of GHGs, clouds, oceans and polar ice sheets... The observed increase [in temperatures] could be largely due to natural variability." The draft of *IPCC AR2* in 1996 said much the same, but various people (like Fred Singer, first director of the US Weather Satellite Center and the University of Maryland, and Frederick Seitz, past president of the National Academy of Sciences, the American Physical Society and President Emeritus of Rockerfeller University) claimed that it was changed before publication by editor Ben Santer, acting on the advice of Senator Tim Wirth, a close confidant of US Vice President Al Gore, whose reputation was increasingly being linked to environmental matters.

Al Gore's scientific training had been one year in the class of Roger Revelle at Harvard (he got Grade D), who had co-authored a study in 1957 about the potential effects of carbon dioxide emissions on the climate. In 1988, when the global warming scare started, Revelle wrote to US Congressman Tim Bates. "Most scientists familiar with the subject are not yet willing to bet that the climate this year is the result of 'greenhouse warming'. As you very well know, climate is highly variable from year to year, and the causes of these variations are not at all well understood. My own personal belief is that we should wait another 10 or 20 years to really be convinced that the greenhouse [effect] is going to be important for human beings, in both positive and negative ways."

This was the caution of a scientist, but the politician Gore could not wait to advance his career and, having been inspired by Revelle's work, he published a book called *Earth in the Balance* in 1992. The same year, in the journal *Cosmos*, Revelle, Fred Singer and Chauncey Starr warned about jumping to conclusions on the subject in an article called *What to Do About Greenhouse Warming: Look Before You Leap*. They wrote, "The scientific base for a greenhouse warming is too uncertain to justify drastic action at this time… The warming reported in the global temperature record since 1880 may thus simply be the escape from this Little Ice Age rather than our entrance into the human greenhouse… If cooling is bad, then warming should be good, it would seem, provided the warming is slow enough so that adjustment is easy and relatively cost-free… With increased atmospheric CO_2, which is after all plant food, plants will grow faster and need less water… Keep in mind also that year-to-year changes at any location are far greater and more rapid than what might be expected from greenhouse warming; and nature, crops and people are already adapted to such changes." Before Singer's death in 2020, he updated his 1997 book *Hot Talk, Cold Science* which repeated the continuing uncertainties about the evidence for AGW and documented the political interference with the scientific process since the 1980s.

While the scientists were cautious, the politicians had lost patience, and so the introduction to *IPCC AR2* made it clear that humans were probably the main cause of climate change. *IPCC AR3*, due out in 2001, would have to do better. The reputation, and even continued existence, of the IPCC was at stake. Tim Wirth had by this time become President of the UN Foundation. The researchers were under extreme pressure to perform, despite inevitable uncertainties in the science. Politicians need certainty to convince the public of the merit of their policies, like Wirth's 1990 'cap and trade' initiative. The ability to predict future climate was still as much in doubt. On page 774 *IPCC AR3* stated: "In climate research and modelling, we should recognise that we are dealing with a coupled non-linear chaotic system. Long-term prediction of future climate states is not possible." The climate is "chaotic" because there are so many factors acting on each other in such a complex way, that they never sit still enough for us to comprehend them and detect predictable patterns.

However, there was evidence that warming was happening, perhaps at unprecedented levels. This was the clinch as far as pinning the blame on humanity was concerned. *IPCC AR3* was most notable for the hockey-stick diagram of historic temperatures (Figure 30), repeated six times in the document, and the controversy it caused. Everything had been going along steadily until the 20th century when the average temperature of the northern hemisphere jumped up sharply and alarmingly.

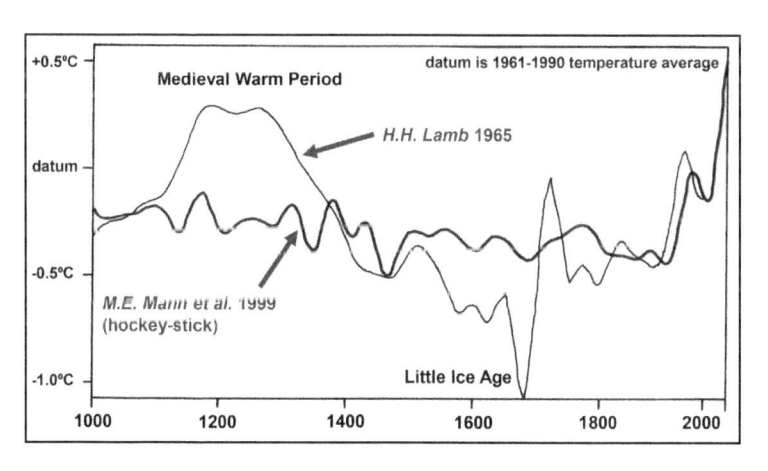

Figure 30 Comparison of Mann's hockey-stick and a previous assessment of historic temperatures by H.H. Lamb.

Levels of CO_2 in the atmosphere had been measured from the tiny bubbles trapped in ice cores for the period up to 1959, when chemical measurements of CO_2 started to be taken at the Mauna Lua Observatory in Hawaii (Figure 23). By joining these two datasets together (Figure 31), a hockey-stick shape was formed, with the period up to 1959 being fairly level (the handle of the stick), if we average the CO_2 between ice age cycles which occur about every 100,000 years. Then there was a marked rise in emissions, assumed to be due to post-1950 industrial development; in other words, on a scale of decades, that is, one ten-thousandth of the horizontal scale unit on which this blip is a scary vertical line. Hey presto, the graphs of increased CO_2 emissions matched that of temperature increases in hockey-stick format, so human emissions must be to blame, and warming was happening at an unprecedented rate.

At first no one seemed to question this publicly, despite the dubious practice of joining together two different data sets on each graph, and the fact that the Medieval Warm Period and Little Ice Age, shown in *IPCC AR1* (Figure 32), had been erased from history. Then, in 2003, Willie Soon and Sallie Baliunas, at

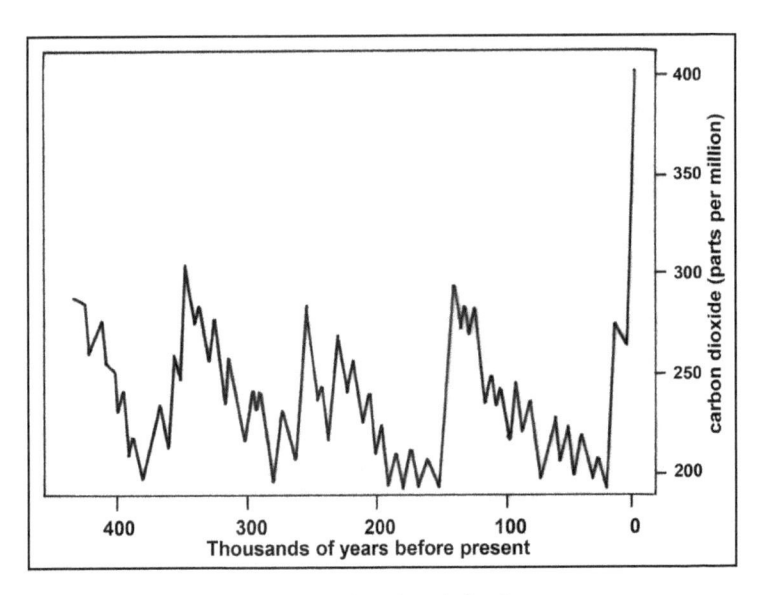

Figure 31 Historic atmospheric carbon dioxide levels.

[Graph showing temperature variation from 1000 to 1900, with labels "Mediaeval Warm Period" and "Little Ice Age"]

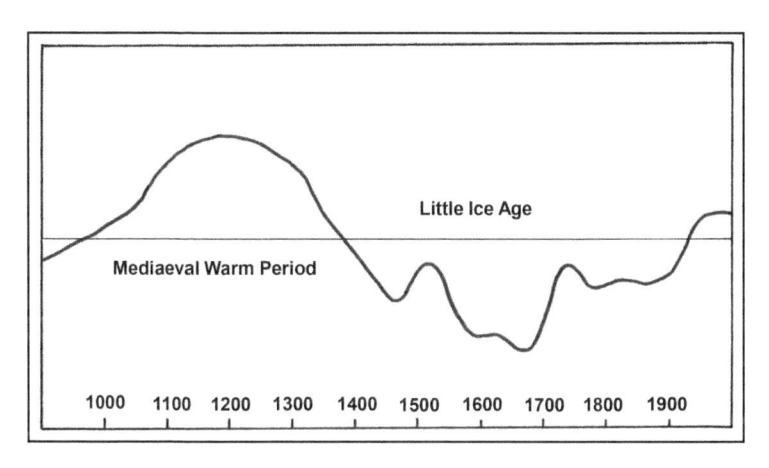

Figure 32 *IPCC AR1* assessment of varying temperatures from the year 1000 to 1990 with the Medieval Warm Period and Little Ice Age clearly shown, and the 20th century only slightly above average.

the Harvard-Smithsonian Center for Astrophysics, reviewed 240 previously published papers to find evidence for the MWP (published in the journal *Climate Change*). It was there, and most of the studies showed that it was warmer than present, and not just in Europe. The late 20th century warming was not unusual after all.

Then, in 2009, Ernst-Georg Beck and colleagues reviewed 170 sources (90,000 measurements) of historic CO_2 before 1960 (Figure 24 in Chapter 9), which showed a slight peak in emissions up to about 380 ppm in the 1930s and '40s, when there was a warming trend. Perhaps at least some of the extra CO_2 in the atmosphere had been due to oceans giving up some of the gas as they warmed, which returned to the general background level of 320 ppm in the cooling period after 1945. Perhaps ice core studies were not sensitive enough to notice such short-term variations.

CLIMATEGATE: THE SECRET'S OUT

Evidence of malpractice by scientists supplying data to the IPCC came on 19 November 2009, when a whistleblower leaked over 1,000 emails from the University of East Anglia, in which the key scientists involved in the formulation of the hockey-stick discussed the subject. The proof came from the horses' mouths. After 1960, the proxy tree ring data showed a fall in temperatures, in contrast with the measured rising temperature trend from the 1970s, indicating that the tree ring data used was suspect. This became known as Climategate, the -gate suffix adopted for subsequent scandals after the resignation of President Richard Nixon due to a cover-up of a 1972 burglary of the Democratic Party HQ at the Watergate office building, approved by him. A further 5,000 emails were released in 2011, described as Climategate 2.0.

There was proof that data was carefully selected to provide the message that the politicians wanted. As dendroclimatologist Rosanne D'Arrigo said at a 2006 US National Academy of Sciences enquiry into the "divergence issue": "You have to pick cherries if you want to make cherry pie". Unhelpful studies would be left out. Phil Jones of the University of East Anglia wrote an email on 8 July 2004: "I can't see either of these papers being in the next IPCC report. Kevin and I will keep them out somehow, even if we have to **redefine what the peer-review literature is**!" Jonathon Overpeck, an IPCC coordinating lead author: "The trick may be to decide on the main message and use that to guide what's included and what is left out."

In September 1999, Keith Briffa, of the University of East Anglia (UEA), had noticed a problem with historic temperature data. He wrote to Michael Mann (Penn State University), Phil Jones (UEA), Tom Karl (director of NOAA's National Climatic Data Center) and Chris Folland (UK Met Office): "I know **there is pressure to present a nice tidy story** as regards 'apparent unprecedented warming in a thousand years or more in the temperature proxy data' but in reality the situation is not quite so simple. We don't have a lot of temperature proxies that come right up to today and those that do (at least a significant number of tree proxies) have some unexpected changes in response that do not match the recent warming. I do not think it wise that the issue be ignored… I believe that the recent warmth was probably matched about 1,000 years ago." Briffa saw that the tree ring data did not show the measured increase in temperature from about 1975 to 1999, and that the MWP had temperatures similar to the end of the 20th century. He was also unhappy about the political "pressure" on them "to present a nice tidy story".

Further dialogue followed between the various men, until Phil Jones emailed Ray Bradley (UMASS Geo Sciences, University of Massachusetts, Amherst), Michael Mann, Malcolm Hughes (Laboratory of Tree-Ring Research, University of Arizona), Keith Briffa and Tim Osborn (UEA) on 16 November 1999 to say: "I've just completed Mike's **nature trick** of adding in the real temperatures to each series for the last 20 years (i.e. from 1981 onwards) and from 1961 for Keith's to **hide the decline**." I have highlighted the key phrases here and in emails below. Jones had hidden the decline in the tree ring data by overlaying other data, and the hockey-stick was born. Sceptics are often accused of inventing conspiracy theories, but six academics from four universities were in at the birth of this "nature trick", by the admission of their own emails.

CRITICISM OF THE POLITICS-LEADING-SCIENCE STRATEGY OF THE IPCC

Various other prominent scientists were unhappy about what was going on. Jeff Severinghaus of the Scripps Institution of Oceanography at the University of California emailed Phil Jones: "If you look at the figure in the attached article in *Science* by Briffa and Osborn, you will note that tree-ring temperature reconstructions are flat from 1950 onwards. I asked Mike Mann about this discrepancy at a meeting recently, and he said he didn't have an explanation. It sounded like it is an embarrassment to the tree ring community that their indicator does not seem to be responding to the pronounced warming of the past 50 years… Personally, I think that the tree ring records should be able to reproduce the instrumental record, as a first test of the validity of this proxy. To me it casts doubt on the integrity of this proxy that it fails this test."

Ed Cook of the Lamont Tree-Ring Lab emailed Keith

Briffa on 30 August 2000: "Here is the Oroko Swamp RCS chronology plot in an attached Word 98 file and actual data values below. It certainly looks **pretty spooky** to me with strong 'Medieval Warm Period' and 'Little Ice Age' signals in it. It's based on substantially more replication than the series in the paper you have to review (hint, hint!)." Rod Savage of the Faculty of Forestry and Environmental Management emailed: "What troubles me even more than the exactness attending chronological estimates is how much **absolute nonsense** – really nothing but **imaginative speculation** – about the environment of the past is being deduced from tree rings and published in dendrochronology [the science of tree rings] journals… As I see it, the peer review process in dendrochronology must be **fundamentally flawed** to allow such publications… your group [should] stop pretending that it knows the answers before it has done the needed research. Again, I am troubled by your group that it shows little humility, **no genuine desire to discover the truth**."

Peter Thorne in the UK Met Office, concerning work for *IPCC AR4* (2007), wrote to Phil Jones: "Observations do not show rising temperatures throughout the tropical troposphere unless you accept one single study and approach, and discount a wealth of others. This is just **downright dangerous**. We need to communicate the uncertainty and be honest. Phil, hopefully we can find time to discuss these further if necessary… I also think **the science is being manipulated to put a political spin on it** which for all our sakes might not be too clever in the long run."

Giorgi Filippo, who contributed to all five assessment reports, in an email dated 11 September 2000 said: "I feel rather uncomfortable about using not only unpublished but also **unreviewed material as the backbone of our conclusions** (or any conclusions)… the rules of IPCC have been softened to the point that in this way **the IPCC is not any more an assessment of published science** (which is its proclaimed goal) but production of results… Essentially, I feel that at this point there are very little rules and **almost anything goes**. I think this will set a dangerous precedent which might undermine the

IPCC credibility, and I am a bit uncomfortable that now nearly everybody seems to think that it is just OK to do this."

Following the Climategate revelations, Tim Ball at the University of Winnipeg joked that perhaps, instead of being in Penn State University, Michael Mann should be in the state pen(itentiary). Mann sued Ball for defamation on three counts (climate activists seem to lack a sense of humour), and initially Ball withdrew through lack of funds. Ball regarded this as a SLAPP tactic. Wikipedia defines a 'strategic lawsuit against public participation' as "a lawsuit that is intended to censor, intimidate and silence critics by burdening them with the cost of a legal defense until they abandon their criticism or opposition. Such lawsuits have been made illegal in many jurisdictions on the grounds that they impede freedom of speech." Ball contends that this is why Mann chose Vancouver as the jurisdiction. However, with crowd funding he re-entered the fray, asking the court to make Mann release the information he had used to compile the hockey-stick. Mann was due in court on 20 February 2017, but did not appear. The Supreme Court of British Columbia ran out of patience in August 2019 and dismissed the case, awarding full legal costs to Ball.

These people knew the flaws in their science. On 12 March 2004 Chick Keller of Los Angeles National Laboratory emailed Richard Somerville, of the Scripps Institution of Oceanography in San Diego, Kevin Trenberth, James Hansen, Phil Jones, Ben Santer and Keith Briffa. He was worried about sceptics he might have to confront: "Their main point is that their counter information hangs together into a logically coherent picture." He summarised the sceptic case as:

1. There is no real fingerprint that distinguishes AGW forcings from others.
2. Models predict warming but the Arctic is hardly warming and the Antarctic is actually cooling slightly.
3. Satellites don't show the same warming as models in the last 30 years.

4. The MWP is likely to have been as warm as present and it was caused by solar changes.
5. Models predict that there should be more warming in the troposphere but satellites do not show this.

MANAGING THE MEDIA AND THE CRITICS

Not only was bad science taking place, but there were attempts to apply pressure on journals and newspapers to present the desired message and avoid proper scrutiny. This is in the last email from Michael Mann on 27 October 2009 before they were leaked, and may have been the final straw that pushed the whistleblower into action: "As we all know, **this isn't about truth at all**; it's about plausibly deniable accusations… Be a bit careful about what information you send to Andy Revkin of *The New York Times* and what emails you copy him in on. He's not as predictable as we'd like." On 23 June that year, Revkin had reported the arrest of NASA Goddard Institute's director James Hansen on a climate demonstration: "Dr Hansen has pushed far beyond the boundaries of the conventional role of scientists, particularly government scientists, in the environmental policy debate".

In 2009 we were also told that William Connelly was tasked with making sure that Wikipedia entries said the "right thing". When those who control YouTube see a post which challenges the orthodox stance on climate change, they insert statements like: "Context: contemporary climate change includes both global warming caused by humans and its impacts on Earth's weather patterns. There have been previous periods of climate change, but the current changes are more rapid than any known events in Earth's history."

Should it be YouTube's function to take sides in any scientific debate, especially as it is impossible for it to be even-handed? It allows Roger Hallam to make a post like *Advice to Young People as they face Annihilation* which incites young people to break the law. Hallam was a founder of Extinction Rebellion and wrote the lecture in one of his twenty visits to prison. He boasts that he was jailed three times before he was 21. He urges "criminal damage" because "The world is a gas chamber... Poisonous gas has been put into that gas chamber for 200 years now, but at a massive rate for the last 30 years, during which people have well known exactly what the consequences would be, and those consequences are now largely locked in. So what we're talking about here is a murder project, a method of murder which involves putting in a gas which is going to destroy the lives and livelihoods of your generation." He says the results will be "war, starvation and rape". Is it YouTube's role to give a platform to serial criminals with unscientific agendas which poison the minds of young people? CO_2 is not a "poison" but a key element of life.

Phil Jones in 2004: "There was some press activity related to this skeptic [Timo Hameranta], but managed to talk the BBC out of doing anything." Jones was worried about how influential a study in the journal *Climate Research* was becoming, which cited the large number of studies that justified the existence of the MWP, and made threats against the co-editor, Chris de Freitas. His email of 3 July 2003 said: "I think that the community should, as Mike M has previously suggested in this eventuality, terminate its involvement with this journal at all levels – reviewing, editing, and submitting, and leave it to wither away into oblivion and disrepute." In other words, if a journal did not tow the orthodox line and, if it applied healthy scientific criticism, steps should be taken to discredit it and promote its failure.

There were three official enquiries into the Climategate affair, which were critical of the University of East Anglia, particularly when it withheld data from information requests, but mildly so, given the time that had passed since the alleged corruption of data. One was chaired by Lord Oxburgh who had financial interests in carbon trading and wind farming, and was vice-chairman of Global Legislators Organisation for a Balanced Environment (Globe International). Steve McIntyre, a former IPCC reviewer, who had been trying to obtain data from the UEA, claimed that the enquiry only reviewed 11 papers from this university, and not critical submissions like the one he had prepared. This was the same Lord Oxburgh who, in

April 2016 along with 12 other members of the UK House of Lords, wrote to the editor of *The Times* saying that if he wished to save the reputation of his newspaper he must stop printing articles which didn't accord with the official orthodoxy on climate change. The establishment was openly trying to stop the free press from being too free.

Matt Ridley's last forthright article on the subject in *The Times* was in December 2016: "So we have an energy policy that has imposed huge costs on the economy, failed to reduce emissions significantly and was either dishonestly or incompetently presented… The costs are actually greater than even the higher-end estimates of damage from climate change. Nobody pays insurance premiums greater than the largest likely loss… In devising its climate-dominated energy policy, government has proceeded as if cost was no object. That is economically irrational, morally wrong and politically foolish. It has needlessly put climate policy on a collision course with public opinion… For rust-belt Americans, just-about-managing Britons, not to mention similar constituencies in Germany, Japan and elsewhere, this is an obvious example of an elite policy that is unfair, costly and futile."

The BBC has officially withdrawn the requirement for editorial balance because it reckons that the issue is beyond question. The only major media company free to be critical of climate change issues seems to be Sky News Australia.

Jonathon Haidt (author of *The Righteous Mind: Why Good People are Divided by Politics and Religion*) reckons that "*The New York Times* is not as reliable as it used to be… A memo went round: 'we need to make race part of every story'." He questions its objectivity. Does it seek to promote "all the truth that's fit to print or is it all the truth that fits our narrative? That's what's happening in some of our social sciences as well." I want newspapers to give me the unadulterated facts in the news pages and keep opinion to the opinion columns, but editors seem compelled to take a stance, from which they cannot extricate themselves, even when there is compelling evidence to show their failings.

Discussion: When people say that there is consensus about a particular subject, or that 'the science is settled' when it rarely is, or that 'the debate is over', should we be suspicious that they wish to avoid a debate about the science and the political policies that have arisen from the 'consensus' science? How free is our free press to criticise governments without losing access to the political contacts who supply information?

SLIDE-CURVES AND CYCLES ARE LESS SCARY THAN HOCKEY-STICKS

Figure 10 in Chapter 3 showed that North Atlantic heat content rose from about 1975 but peaked in 2007 and started to get cooler. This coincides with Arctic sea ice reducing from about 1979 to 2007, with Figure 11 showing no trend, or if anything a slightly upwards trend, in the period 2007 to 2019. It seems to make sense that floating ice will be more sensitive to the temperature of the water it is on than the air above it, which we are told by the IPCC has only risen by 1°C in a century. If we want to get ice cubes out of their mould quickly, we float them in warm water rather than just leaving them at room temperature. The correlation between the rise and fall of the Atlantic water temperature and the Arctic sea ice is not perfect because it takes time for ocean currents to circulate, but it is pretty good. The Arctic is also affected by North Pacific ocean currents, although less so because of the narrowness and shallowness of the Bering Strait. These also vary cyclically.

In Figure 33 I joined together the mean estimate by *Vinnikov et al.* 1980 of Arctic sea ice from the 1920s to 1978 and the satellite measurement from 1979 to 2019 PIOMAS data. The former is probably less accurate because of the less sophisticated means of measurement, and its calibration to fit on the same scale as the satellite data can only be approximate, but the main point is that it shows the cyclical trend of the ice from high in the 1920s, to a low point in the 1950s, to a high point around 1980, to a low point around 2010. In 1959, the nuclear submarine USS Skate surfaced at the North Pole when

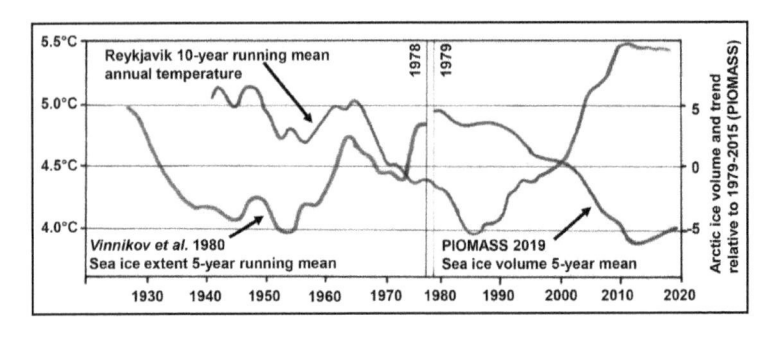

Figure 33 Comparison of ice extent and Iceland temperatures.

ice was thin (the newsreel is on YouTube). Since *IPCC AR2*, the UN has only considered the reducing line from 1979.

The temperature of the ocean affects the climate of coastal locations. I know this because palm trees grow on the southwest coast of Scotland which is kept mild in winter by the Gulf Stream/North Atlantic Drift. So in Figure 33 I took the mean temperature of Reykjavik from 1940 to 2019 which is a pretty good mirror image of the ice extent, with warming in 1940 to cooling in the early 1980s to a peak in warming around 2010. The land surface temperature in 2010 (mean 5.0°C) seems to be about half a degree warmer than the 1940 peak, but the records I found only started then, and it may have been a bit warmer in the 1930s. Even if it wasn't, Reykjavik has only experienced less than half a degree centigrade rise in temperature in the last century. That doesn't seem anything to worry about, and Chick Keller was right to say that "the Arctic is hardly warming". It might even be worth celebrating if you are in a country with an average temperature of only 4.5°C. Chick might agree that my "counter information hangs together into a logically coherent picture".

Figure 33 only has one-and-a-half cycles, but sailors have been measuring the temperature of the water on the busy North Atlantic shipping route since the 1860s. The means of measurement were not very sophisticated. A bucket would be dropped into the water and its temperature taken with a thermometer. Later, the water would be pumped directly into the ship, but it could still be contaminated by the heat of the ship's engines. Even with these caveats in mind, we can draw the conclusion that the Atlantic Multidecadel Oscillation is a cyclical phenomenon that goes up and down in temperature about every 60 years (its frequency). Figure 11 suggests that the drop in Arctic ice has paused and Figure 33 suggests that perhaps we might see a significant increasing trend in ice cover by 2030. We will have to wait and see.

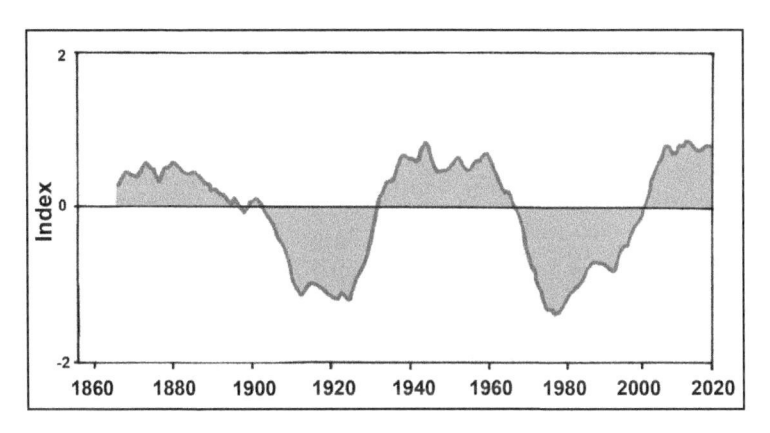

Figure 34 North Atlantic ocean temperature anomaly 1860s to 2020.

Russian scientists are less influenced by the American and European ones advising the UN, and may be more inclined to accept the evidence on the cyclical nature of Arctic ice. Perhaps, because of the advice he was being given about Arctic ports possibly icing up more in winter, Vladimir Putin hedged his bets by ensuring that he had a warm water port at Sevastopol when his troops took the city in 2014. He had other 'patriotic' reasons for taking back Crimea, but this may have clinched it.

THE MEDIEVAL WARM PERIOD AND THE LITTLE ICE AGE
Data from Soon and Baliunas and D E Koelle is plotted on Figure 35 to show how the climate has varied since the MWP. The striking thing for me is how humanity prospered during warm periods, and suffered in the LIA due to colder and probably drier conditions. Historical events seem to confirm the data. For just over 400 years

in the MWP, the Vikings survived in Greenland and the English grew vineyards, which have only been commercially viable since the 1980s in the present warm period. The Vikings could sail round Greenland, a feat which is still not possible today, and their dead are buried in ground which is still in permafrost that may take another century of warming to start to thaw out. The MWP was also when the great Gothic cathedrals were built throughout Europe due to the prosperity from the 12th to 14th centuries.

There are records of English vineyards from soon after the Norman Conquest in 1066, with 42 locations recorded in the Domesday Book in 1086. By the time of Henry VIII (ruled 1509–1547) there were 11 vineyards owned by the Crown, 67 by noblemen and 52 by the Church. The Dissolution of the Monasteries in 1536 probably ended their vineyards, and the colder conditions of the 17th century caused the demise of them all. Attempts to cultivate vines from the 17th to mid-20th centuries were unsuccessful. The Nyetimber Valley Vineyard in the south of England planted vines in 1988 and was the site in medieval times of what was then known as the Nitimbreha Vineyard, as recorded in the Domesday Book.

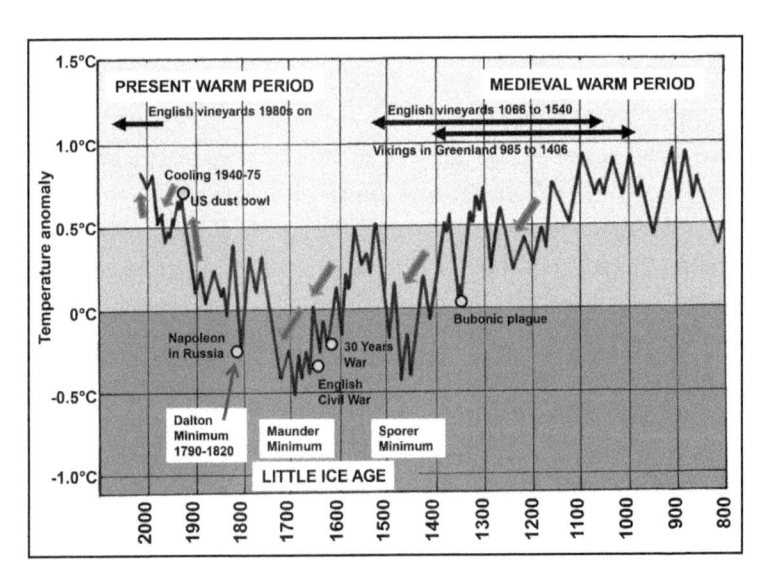

Figure 35 Temperature variation since the Medieval Warm Period.

The Black Death (1347–49) occurred at a cold period just before the LIA started. People can catch a virulent disease at any time, but if a colder period affects food supplies, they will be weaker to resist it and more will die. During the coldest part of the LIA, the Maunder Minimum, historian Geoffrey Parker (*Global Crisis: War, Climate Change and Catastrophe in the Seventeenth Century*) calculated that around a third of Europeans died from cold, starvation, disease and civil conflict. Upland areas which had marginal farmland were abandoned for coastal plains. The Dalton Minimum was a brief return to cold conditions and was unfortunate for Napoleon's Russian campaign. His military strength and tactics were much superior, but the summers up to 1812 did not produce sufficient food for an army of 610,000 men. They were in such poor physical condition that Napoleon could only muster 135,000 troops for the Battle of Borodino, before he entered Moscow in October. With the harsh Russian winter setting in, he had no option but to retreat, but only 100,000 men made it back to France alive.

The Dalton Minimum did have benefits for Russia and Canada, which could slaughter millions of furry wild beasties to sell their coats to freezing Europeans. Britain had invented ways of mass-producing cotton clothing much cheaper than on the Continent, much to Napoleon's annoyance, and the money gained from this helped fund the rest of the Industrial Revolution, such that by the end of the 19th century, British people owned 44% of the world's capital, to invest in infrastructure in their expanding Empire and beyond.

In my own family history research, on my mother's side, I could trace relatives in central Scotland back to the 17th century. Life in the LIA was hard for farmers with my great-great-great-great-great-great-grandfather dying at age 40. His son moved into the local town, became a merchant in the weaving trade and lived to age 60, leaving farmer cousins to face the violent hail storm of 27th June 1733 which flattened crops and drowned farm animals during floods that followed. Frosts as early as September some years ruined harvests.

In the town, 60 people died of pleuritic fever in the winter of 1733, but linen from local flax was eventually replaced by cotton from the American colonies, and the town prospered. The weavers would not give any thought, even if they knew, that the cotton was farmed by slaves, because they were fully absorbed in their own fight for survival. Indeed the Bible (and the Quran) made it clear that slavery was an integral part of pre-industrial 'civilisations' and did not condemn it. It was only with the wealth and machinery brought about by the power of coal that people could afford to entertain the moral argument for abolition which, it is no coincidence, happened first in the birthplace of the Industrial Revolution. As Figure 26 showed, coal use grew towards the end of the 20th century, and will not reduce by 2040, as countries use this fuel, gas and oil to release their people from the chains of absolute poverty, the way my relatives benefitted in the late-18th and 19th centuries, most of them by then living into their 70s. The last thing poor countries today are interested in is a rich-man's theory about an atmospheric trace gas causing climate chaos, even if it turns out to be true.

The Little Ice Age in Europe, with its cooler and drier climate, coincided with periods of drought in sub-Saharan Africa. Europeans had taken advantage of an indigenous slave trade and, according to John Reader in his 1997 book *Africa: A Biography of the Continent*, "Portuguese records from the 16th to the 19th centuries reveal a direct correlation between ecological stress, warfare, and the export of refugees as slaves. Drought, famine and disease were a formative component of the disruption and chaos which carried the slaving frontier from the Atlantic coast to the heart of the continent, where it collided with a wave of similar proportions that had advanced from the Indian Ocean coast. Droughts of devastating intensity afflicted the populations of west-central Africa during the 1570s and '80s, the 1640s and '50s, the 1710s and '20s and almost continuously from 1784 to 1795.

"Five thousand people had occupied the Kongo kingdom's capital in 1774, for instance, but only 100 were there to be counted in the 1790s. Of the remainder, some had left the town to search for food in the surrounding countryside, others had starved; many had been sold into slavery." Most transactions involved a handful of people at a time as the survivors sold their kinsfolk in an attempt to survive. In the West we assume that the inhumane slave trade ended at the beginning of the 19th century (Britain 1807, USA 1808, Netherlands 1814, France 1818 and Brazil 1830) but slavery continued within Africa and in certain Muslim countries for much longer, with 1,650,000 more slaves sold across the Sahara and the Red Sea. Saudi Arabia only abolished the legal status of slavery in 1962. In contrast to the Americas, there are few black people of sub-Saharan origin in Arab countries today because the slaves were castrated in transit.

Many scientists thought that the cooling from about 1940 to 1975 was the harbinger of a return to LIA conditions, which was understandable since this had been the default position for 500 years.

ARCTIC ICE AS A MEASURE OF HOLOCENE CLIMATE

Stein et al. 2017 determined past Arctic Sea ice levels by observing the amount of phytoplankton which sank to the bottom of the Arctic Ocean in historic layers of sediment. The more phytoplankton, the warmer the ocean was at the time, and the less ice. They surveyed two locations, one more affected by Pacific Ocean currents and the other more by Atlantic ones (the solid and dotted lines in Figure 36).

It is extremely difficult to calibrate historic biomarkers like this so, rather than guestimating the actual temperature differences between eras, the vertical axis is split into three bands: 'reduced' when there was significantly less Arctic ice than present; 'seasonal' when the ice level varied significantly between summer and winter as happens just now; and 'perennial' when the ice stayed constant all year round because of cold conditions. The last of these only occurred during the LIA, the coldest period in the last 10,000 years. In 2014 Geologist Don J. Easterbrook at Western Washington University obtained similar results from studies of Greenland ice cores.

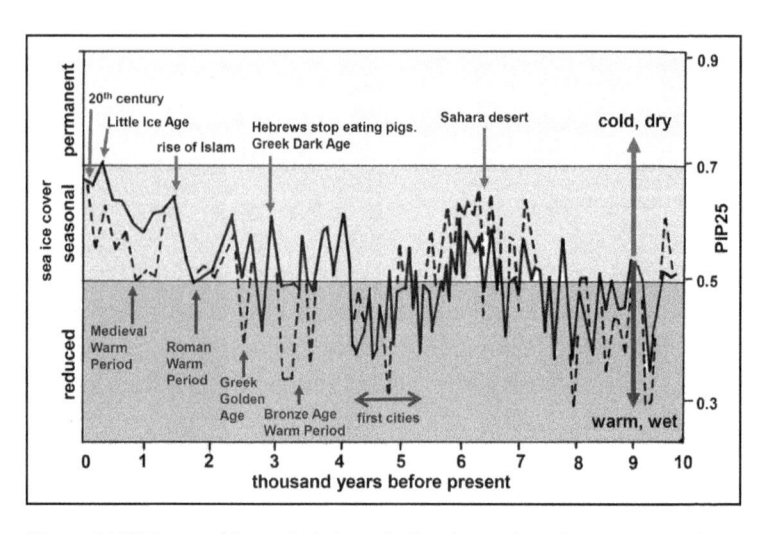

Figure 36 Holocene (the period since the last ice age) sea ice cover variations.

HISTORY CONFIRMS THE PROXY DATA ON CLIMATE CHANGE

Tim Lenton, in his 2011 book *Revolutions that Made the Earth*, noted the changes in civilisations due to changing climate. "As people built and expanded agriculturally fuelled empires, they had an increasing impact on their environments, but they were also subject to natural changes in climate. Although the Holocene used to be portrayed as having an unusually stable climate, and life in the high northern latitudes was certainly less tumultuous than during the last ice age, there were substantial climate changes in the tropics. Sedentary civilisations based on agriculture were (and still are) more sensitive to climate change than mobile foraging societies, and climate change has been implicated in the rise and fall of some civilisations… Around 5,000 years ago, a large-scale climate change, involving the drying and collapse of savannah ecosystems in the Sahara, brought a 200-year period of drought to the Middle East. This has been implicated in the collapse of Sumerian Late Uruk society. With the return of wetter conditions, various class-based and politically centralised societies flourished in the Old World, including the Akkadian Empire in Sumer, but they too collapsed in a further interval of drought around 4,200 years ago.

"In the New World, the collapse of the Mochica (coastal Peru, 1,500 years ago), Classic Maya (Yucatan Peninsula, 1,200 years ago) and Tiwanaku (Bolivia–Peru, 1,000 years ago) empires have been linked to sustained (natural) drought intervals. In more recent human history, correlations have been noted between intervals of cooler climate, poor agricultural production, war and population decline." [*Zhang et al.* 2007]

The first cities arose in Mesopotamia, the Indus Valley and China during a warm period, as did the earliest major Middle Eastern Bronze Age civilisations: the Egyptians, the Hittites, the Assyrians and the Babylonians. They generally lived in peace with each other because their prosperity depended on trade. Bronze is made from copper and tin. The name Cyprus means copper, and the tin came from distant places like Afghanistan and Cornwall. Around 1200BCE the climate changed rapidly to become colder and drier. Crops failed some years, malnutrition and disease became prevalent, civil strife increased and there were mass migrations. The Egyptian, Assyrian and Babylonian Empires shrank back to their core territories close to the Nile, Tigris and Euphrates, and the Hittite Empire completely disappeared within a few generations.

In 1177BCE Egyptian sources described the occupation of the fertile coastal plains of present-day Israel by invading Sea Peoples. They, and the Canaanite tribes which inhabited the hill country between the coastal plain and the River Jordan, were largely left alone by major powers for four centuries to develop their own cultures, the main differentiating feature of the hill tribes being their rejection of pig meat as a source of food. About 4,500 people lived in 250 settlements according to Israel Finkelstein and Neil Asher Silberman in their book *The Bible Unearthed*. During periodic skirmishes between the various small tribes living between the Nile and Euphrates valleys, each dreamed about being the sole master of this power vacuum. The hill tribes developed a mythology that became the basis of the Jewish religion in which they cast themselves as

God's chosen people who had the right to claim all of the land between the two rivers, and expel anyone else who lived there. The tribes coalesced into two kingdoms, Israel in the north and Judah in the south. Without the advent of a cold period that shrank the influence of the large powers, the Israelite/Hebrew/Jewish people might have been absorbed into another culture, and Judaism might not have been founded.

In his book *Ancient Greece*, Thomas R Martin noted that "The centres of Mycenaean civilisation were destroyed in the period from about 1200BC to 1000BC as part of widespread turmoil throughout the eastern Mediterranean region. The descendants of the Greeks who survived these catastrophes eventually revived Greek civilisation after the Dark Age (1000–750BC)… centuries of economic devastation, population decline, and political vacuum… Only Athens seems to have escaped wholesale disaster," as opposed to other parts of Greece, where "they wandered abroad in search of new territory to settle".

The Romans also prospered during a warm period, with the trade in agriculture and commodities, mostly around the Mediterranean Sea, focused on its centre, Rome, until Constantine set up the competing capital at Byzantium, renaming it Constantinople (now Istanbul which is its Muslim name). A 1999 study by Ian Meadows of Northampton County Council and Tony Brown of Exeter University identified seven Roman vineyards in England with an estimate of 250 square miles of land under vineculture. Many reasons have been considered for the decline and fall of the Roman Empire, but it suffered when temperatures cooled from the end of the 4th century CE. The Huns from the Steppes of modern-day Russia moved westwards towards the Atlantic rains. The mobile, horse-based culture pushed aside the tribes which lived east of the Rhine, and forced them into Roman territory. A Rome weakened by poor harvests had little power to resist their migrations, and the indigenous British population had to accept greater numbers of Anglo Saxons into their communities, and the adoption of English as their language. The incursions were often peaceful because wars are costly, but the waning wealth of Rome was plundered on the three occasions that it was sacked, by the Visigoths in 410, the Vandals in 455 and the Ostragoths in 546.

The Eastern Empire withstood the onslaught of migrants better, although by 541 plague had spread from Alexandria on ships carrying grain to Constantinople. For most of the 6th century the Eastern Roman Empire and the Persian Empire fought over scarcer resources, until the final war from 626 to 630, after which they were both utterly exhausted. The Roman army went on strike because it was not being paid. Ominously for Constantinople and Persia, Mohamed died in 632, and a new force and ideology was set to fill the power vacuum. (The single surviving contemporary account of him, a Christian one, still had him alive in 634, with Muslim accounts of his life only being written over a century later.) Arabs, with their camel trains, had been merchants and often mercenaries when war took place, not being fussy whether they joined the Romans or the Persians. They did not live in the big cities of the empires which were occasionally stricken by plague, and they possessed the appropriate skills and mobility to replace the two fading powers without much difficulty. By 636 an Arab army had taken Damascus and Homs, and by 642 Alexandria was under its control. An invasion of the Persian Empire took place in 649. Much of modern-day Turkey was conquered by 674. They reached Spain in 711.

CONFIRMATORY EVIDENCE FOR HISTORIC CLIMATE CHANGE

Such historical events seem to bear out the Stein study, but there are many other studies which confirm its results. Steven Emslie (University of North Carolina Wilmington – *GeoScienceWorld* Sept 2020) found penguin remains from around the year 1200 below melting ice at the edge of the Ross Sea in Antarctica. Fred Goldberg calculated that the Bronze Age Warm Period was about 3.05°C warmer than the year 1800, the Roman Warm Period about 2.3°C warmer, the MWP about 1.3°C warmer. *Margaritelli et al.* 2020 in *Scientific Reports* identified the Roman and Medieval Warm Periods in Mediterranean sea deposits. *Mangerud and Svendsen* 2018, *Bartels et al.* 2018, and *McFarlin et al.* 2018 identified a Holocene Thermal Maximum (the warmest period since the last Ice Age) about 9,000 years ago, when Greenland may have been 4°C to 7°C warmer than present. "Shallow marine molluscs that are today extinct close to Svalbard in the Hinlopen Strait, because of the cold climate, are found in deposits there dating to the early Holocene. The most warm-demanding species found, Zirfaea crispate, currently has a northern limit 1,000 km farther south, indicating that August temperatures on Svalbard were 6°C warmer at around 10,200BP to 9,200BP [before present], when this species lived there… One single specimen of Mytilus is dated to 900BP, suggesting a short-lived warm period during the Medieval Warm Period of northern Europe."

A team at the Institute of Geology and Geophysics in Beijing published a study in *Nature Communications* in September 2019 which traced ancient Chinese civilisations and found a correlation with 500-year oscillations of monsoon climate, probably caused by solar cycles. They analysed the pollen record from the annually laminated Maar Lake in Northeast China to obtain a proxy of monsoon climate, together with 627 radiocarbon dates from archaeological sites as a proxy of human activity. "The warm-humid/cold-dry phases of monsoon cycles correspond closely to the intensification and weakening of human activity and the flourishing and decline of prehistoric cultures." The warmer the climate, the more prosperous the civilisation was in terms of grain cultivation, animal domestication and human settlement. Previous studies of civilisations could only consider the last 3,600 years, since writing was invented. Lead author Xu Deke concluded that "there is a natural constraint on human efforts".

Societies started to collapse when cooling took place, and neither culture nor political systems could sustain them. If the solar cyclical activity continued as before, Xu postulated that over the next few decades the Earth would enter 25 years of cooling, although increased greenhouse gases could slow the temperature drop. "The most effective counter-measure is science and technology. We are in a much more capable position than our ancestors with the help of technology and machines in the face of global cooling, but preparation must start now."

Goodness me. A scientist is telling us to start preparing now for the cold. Figure 35 shows that the LIA was the coldest period in the Holocene, and the 21st century is only about 1°C warmer than this. Perhaps we need to hope that the AGW theory is correct and more emissions will raise the temperature closer to the more benign conditions that led to previous periods of successful human civilisation. Or perhaps we should ignore all such predictions, accept whatever climate we get, and do what we do best – adapt.

THE NASA GODDARD INSTITITE UNDER JAMES HANSEN

We previously came across James Hansen in Chapter 3 making scary climate predictions that did not transpire. We then saw him earlier in this chapter as a correspondent in Climategate emails. He was Director of the NASA Goddard Institute from 1981 to 2013. In December 1989 he declared: "By the year 2050 we're going to have tremendous climate changes far outside what man has ever experienced." He gave testimony to Congress that temperature increases would be between 4°F and 9°F (2.2°C and 4°C). If you look at his profile in Wikipedia you will see a photo of him being arrested for breach of the peace outside the White House on a climate demonstration, the third time he had been arrested in similar circumstances. Hansen has always been an activist rather than a dispassionate scientist, and was allowed to flaunt civil service rules that prevented senior bureaucrats taking part in political activities.

He has also earned a large amount of money lecturing on climate change. To quote geologist Ian Plimer (University of Melbourne): "In a period of five years, Hansen earned in outside income between $1.47 million and $2.67 million, in addition to his basic salary as a government employee of $180,000. Under the terms of contract governing that salary, Hansen is forbidden from privately benefitting from public office and from taking money for activities related to his taxpayer-funded employment." No one in political circles seemed to have called him to account.

It is difficult to see how the NASA Goddard Institute could have had a balanced approach under his direction, and Tony Heller is a notable critic of it. He spent 20 years as a microprocessor designer at Motorola, where he prided himself in the robustness of his due diligence, since mistakes in software could have cost his employers millions of dollars. He prepared software for weather and climate models for the National Center for Atmospheric Research, carried out geothermal energy research at Los Alamos, produced imaging systems for drones for use commercially and by the Defense Department, worked on Google's virtual reality system and engineered remote surgical equipment. Working at Los Alamos in 1980, his boss introduced him to anthropogenic global warming and he was a "true believer" until 2003 when he started to question the science. He was appalled by the sloppy practice within the Goddard Institute which seemed to be a law unto itself, like its director. Such unprofessionalism could not have happened under the commercial scrutiny of the private sector organisations that he had been used to.

We have already seen that both historic and current temperature records have accuracy problems for various reasons such that there is a big margin of error and a certain latitude of possible interpretation. Heller believes that the Goddard Institute has exploited this to tell a story that politicians like Al Gore would like. For example, Figure 7 showed a global cooling period from about 1940 to 1975. In Figure 37 I show the data he took from the IPCC 1990 graph of global temperatures and superimposed the data 'recalibrated' by the Goddard Institute in 2016. The cooling period had disappeared, just like the MWP and LIA had gone from Michael Mann's hockey-stick, and the 1975–98 warming trend started from a higher point.

Heller then compared the Goddard Institute's US temperature graphs from before the year 2000 and in 2011 (Figure 38). In the former, the 1930s, the time of the Midwest American Dust Bowl, the temperature peaked and was warmer than the late 1990s,

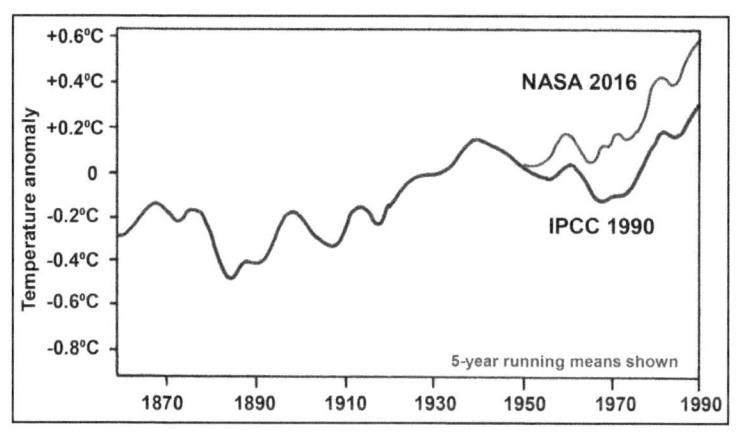

Figure 37 Comparison of global temperature graphs, IPCC 1990 and NASA Goddard Institute 2016.

but in the 2011 version this was reversed. Don J. Easterbrook at Western Washington University also noted that Greenland temperatures in the 1930s were warmer than in the 1990s in his paper in *Kalte Sonne* of 12 October 2014.

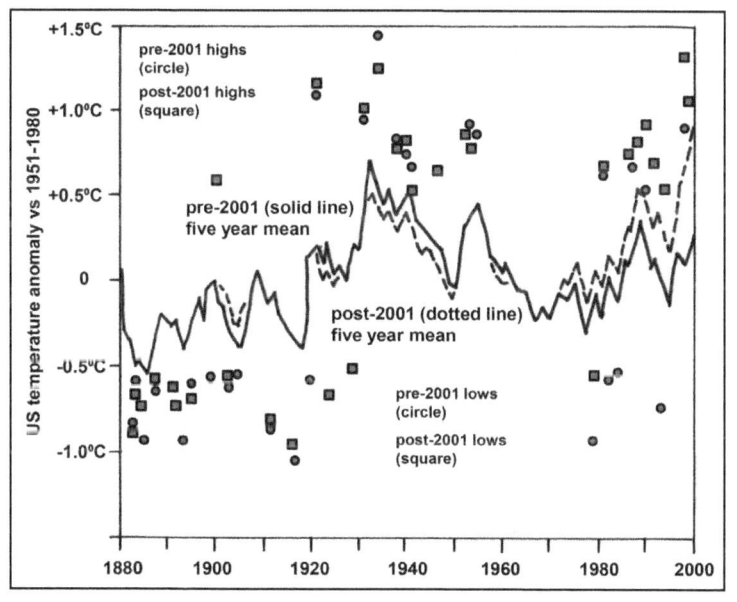

Figure 38 Comparison between NASA pre- and post-2001 (dotted line) US temperature graphs.

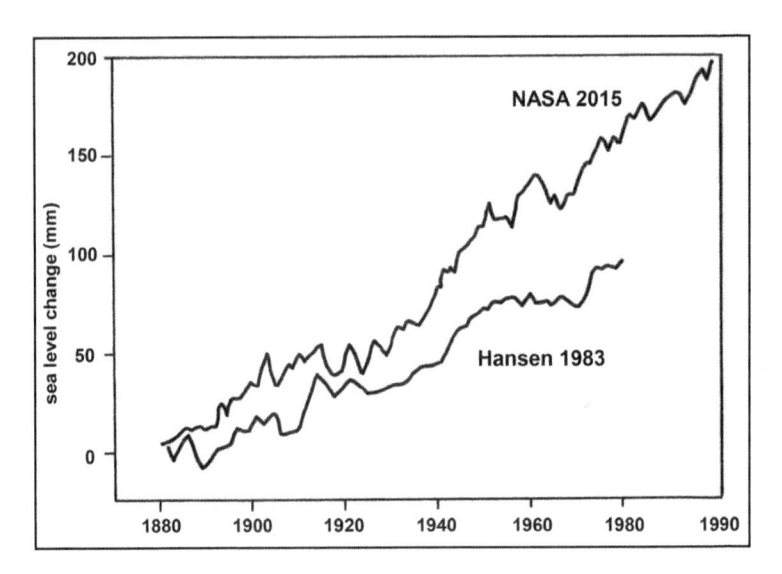

Figure 39 Sea level change comparison.

Heller also noticed that NASA had adjusted the sea level data since the 1980s (Figure 39). He plotted their 2015 graph on a Hansen record from 1983, data which he says was confirmed by others like Vivien Gornitz of NASA. Again the orthodoxy in 1990, as referenced in *IPCC AR1* was "From examinations of both composite regional and global curves and individual tide gauge records, there is no convincing evidence of an acceleration in global sea level rise during the twentieth century."

Heller is a sceptic, although there are many who regard him as an irresponsible conspiracy theorist. What these graphs show me is that small changes in calibration or interpretation, or variations in inevitably imperfect data collection, can lead to significantly different outcomes over time.

WHO AND WHAT TO BELIEVE?

Those of us in Western democracies are living in the most benign societies that have so far existed. That is not to say that everything in the garden is rosy but I would rather live in an imperfect society with capitalism (with an inevitable spectrum of individual wealth), democracy (with semi-scrupulous politicians), freedom of speech (with elements of political correctness and cancel culture), a free press (even if titles pander to the prejudices of their readership to stay in business), tradition and slow change (even if flawed traditions like racism and sexism take time to eradicate), an independent judiciary (if subject to political appointees in America) and materialistic, free-choice consumerism (even if it leads to modern health problems like obesity). An oppressive, dictatorial state is far worse.

In various Meetup groups I have come across many young people with all sorts of fears about modern life which seem to me out of proportion, given the comparatively more sexist, racist, bigoted and materially impoverished environment I experienced as a young person, with fears of the Cold War and a procession of economic recessions. Chief among the new fears is climate change, where there seems to be a desire to believe the worst and find attraction in the doom-mongers, with no time for anyone who dares to try to moderate the alarm. It is sad to witness such needless anxiety.

Should we believe Tony Heller, whose software design at Motorola and Los Alamos contributed to human progress, or James Hansen, a lifelong academic who might not last long under the harsh critique of the commercial world, and whose alarmism has led to highly flawed energy policy?

I have criticised those associated with Climategate, but they believed that the end justified the means, and did what they thought was right. I am sure Phil Jones was being earnest when he made the following unscientific remark in an email: "There is no way the MWP (whenever it was) was as warm globally as the last 20 years. There is also no way a whole decade in the LIA period was more than 1 deg C on a global basis cooler than the 1965–90 mean. **This is all gut feeling, no science**, but years of experience of dealing with global scales and variability. Must go to Florence now."

Steve McIntyre is a notable 'debunker' of the hockey-stick (climateaudit.org), Michael Mann wrote a book to defend it (*The Hockey Stick and the Climate Wars: Dispatches from the Front Lines* 2014) and some others have found proxy data to support the contention that temperatures were relatively stable before the modern era. But, instead of polarisation, why can we not have a presentation of all the pertinent data and an informed analysis by an appropriate broadcaster in terms that intelligent people can understand? The subject is too important for us just to take certain people's word for it, in a subjective and partisan way.

I can watch commentators on YouTube and judge their personalities to assess their credibility. Do they seem well informed and think for themselves, or do they just roll out a standard message? Do they have a sense of humour that lets them be self-depricating and not take themselves too seriously, such that they are open-minded? Do they expose themselves to informed criticism or do they restrict themselves to young or sympathetic audiences with limited alternative knowledge?

In a 2010 lecture (YouTube: *Mann's Hockey Stick "Whitewash"*) McIntyre said, "People are far too quick to yell fraud at the other side, and I think that such language is both self-indulgent and counterproductive. I don't apply these labels myself, I don't permit them in climateaudit, and don't believe they serve any purpose… Any point you make should be done on the facts, not the adjectives." In his 2021 book *The New Climate War: the Fight to Take Back Our Planet*, Mann used the words 'deny' or 'denier' 101 times, and accused those who contest the need to "avert catastrophic global climate change" as tapping into "reptilian brain impulses". He must mean me. In my opinion, the use of insults like this, as an alternative to cool, rational discourse, is a sure sign of a lost cause. It seems that people like Mann, Hansen, Jones and Anderson were appointed to influential public posts due to their enthusiasm for the subject, but that they became ideologues rather than open-minded scientists, and that public organisations lack the ability to refresh themselves with contrary and new ideas in the way that commercial firms must to respond to evolving market situations.

I like those who were converts to climate alarm but have seen the light, because they observe the issues clearly from the inside, dishing the dirt. Michael Shellenberger co-founded the

US Democratic New Apollo Project which was later adopted by President Obama, with $90 billion spent on stimulus money "for efficiency, renewables, electric cars and other technologies... Stimulus money wasn't evenly distributed but rather clustered around donors to President Obama and the Democratic Party... The loans for electric car companies like Tesla and Fisker, each of which received nearly half-billion dollars, had no performance measures... The most famous of the green investments was when DOE gave $573 million to a solar company called Solyndra, 35% of which was owned by a billionaire donor and fundraising bundler for Obama, George Kaiser... The people who benefited the most from the green stimulus were billionaires, including Elon Musk, John Doerr, George Kaiser, Vinod Khosla, Ted Turner, Pat Stryker and Paul Tudor Jones."

Shellenberger now supports a policy to minimise emissions based on proven technologies like hydro, nuclear and gas, and showed that fossil fuel companies had collaborated with 'green' organisations which opposed nuclear plants and hydro dams because they had a common interest, one based on environmental dogma and one on commercial motives. "In February 2012, the new executive director of the Sierra Club went to *Time* with a confession: his organisation had accepted more than $25 million from natural gas investor and fracking pioneer Aubrey McClendon." Tom Steyer has funded 350.org, and his company Farallon Capital Management invests in fossil fuels.

The National Resources Defense Council and the Environmental Defense Fund are also funded by fossil fuel billionaires. "The two largest US environmental organisations, EDF and NRDC, have a combined annual budget of about $384 million compared to the mere $13 million of the two largest climate skeptic groups, Competitive Enterprise Institute and Heartland Institute." Activists claim that the fossil fuel industry funds unscrupulous sceptic organisations, but it is happy to sit back and gain from our inevitable continued reliance on oil and gas, obtain public subsidies when they can for otherwise unprofitable renewables and make selective donations to green organisations for PR purposes when their interests align.

Fellow of the American Association for the Advancement of Science, and previously undersecretary for science in Obama's Department of Energy, Steve Koonin, after he had reviewed the science in 2014 for the first time, compared the AAAS's high level statement on climate change with what he would have said.

American Association for the Advancement of Science: "Our nation, our states, our cities and our towns face an urgent problem: climate change. Americans are already feeling its effects and will continue to do so in the coming decades. Rising temperatures will impact farmers in their fields and transit riders in cities. Across the country, extreme weather events such as hurricanes, floods, wildfires and drought are occurring with greater frequency and intensity. While these problems pose numerous risks to society and the planet, undoubtedly the biggest risk would be to do nothing. Science tells us that the sooner we respond to climate change, the lower the risks and the costs will be in the future."

Steve Koonin: "The earth has warmed during the past century, partly because of natural phenomena and partly in response to growing human influences. These human influences (most importantly the accumulation of CO_2 from burning fossil fuels) exert a physically small effect on the complex climate system. Unfortunately, our limited observations and understanding are insufficient to usefully quantify either how the climate will respond to human influences or how it varies naturally. However, even as human influences have increased almost fivefold since 1950 and the globe has modestly warmed, most severe weather phenomena remain within past variability. Projections of future climate and weather events rely on models demonstrably unfit for purpose." Koonin's analysis of current climate change science is in his 2021 book *Unsettled: What Climate Science Tells Us, What It Doesn't, and Why It Matters.*

After the Climategate 2.0 leak, Roy Spencer (ex-NASA and now University of Alabama in Huntsville) wrote in November 2011: "We are all familiar with competing experts in a trial who have diametrically opposed opinions on some matter, even given the same evidence. This happens in science all the time. Even if we have perfect measurements of nature, scientists can still come to different conclusions about what those measurements mean in terms of cause and effect. Biases on the part of scientists inevitably influence their opinions. The formation of a hypothesis of how nature works is always biased by a scientist's world view and limited amount of knowledge, as well as the limited availability of research funding from a government that has biased policy interests to preserve.

"Admittedly, the existence of bias in scientific research – which is always present – does not mean the research is necessarily wrong. But as I often remind people, it's much easier to be wrong than right in science. This is because, while the physical world works in only one way, we can dream up a myriad of ways by which we think it works. And they can't all be correct. So, bias ends up being the enemy of the search for scientific truth because it keeps us from entertaining alternative hypotheses for how the physical world works. It increases the likelihood that our conclusions are wrong.

"In the case of global warming research, the alternative (non-consensus) hypothesis that some or most of the climate change we have observed is natural is the one that the IPCC must avoid at all cost. This is why the hockey-stick was so prized: it was hailed as evidence that humans, not nature, rule over climate change. The Climategate 2.0 emails show how entrenched this bias has become among the handful of scientists who have been the most willing participants and supporters of the cause. These scientists only rose to the top because they were willing to actively promote the IPCC's message within their particular fields of research. Unfortunately, there is no way to 'fix' the IPCC, and there never was. The reason is that its formation over 20 years ago was to support political and energy policy goals, not to search for scientific truth. I know this not only because one of the first IPCC directors told me so, but also because it is the way the IPCC leadership behaves. If you disagree with their interpretation of climate change, you are left out of the IPCC process. They ignore or fight against any evidence which does not support their policy-driven mission, even to the point of pressuring scientific journals not to publish papers which might hurt the IPCC's efforts.

"I believe that most of the hundreds of scientists supporting the IPCC's efforts are just playing along, assured of continued funding. In my experience, they are either: (1) true believers in the cause; (2) think we need to get away from using fossil fuels anyway; or (3) rationalise their involvement based upon the non-zero chance of catastrophic climate change… I hope I am correct that most climate change we have experienced is natural. But I also know that 'hoping' doesn't make it so. If I had new scientific evidence that human-caused climate change really was a threat to life on Earth, I would publish it. It would sure be easier to publish than evidence against. But from everything I've seen, I still think nature probably rules, and that humans (as part of nature) also have some unknown influence on climate."

I like what Roy Spencer says. It oozes common sense. When I emailed him with queries, having never had any contact with him before, he answered the same day. My unsolicited correspondence with sceptic Matt Ridley and Hans Rosling's son Ola were also met with kind responses. These nice people just seem level-headed and refreshingly open-minded. In contrast, I have found advocates of the pessimistic orthodoxy of climate change to be generally humourless and intransigent. I like the former much better.

Discussion: Since there seem to be no completely independent sources of information, how can we separate hard scientific facts from those which have been influenced by politics and other ideologies? Are we more likely to believe people with the same outlook as ourselves and because we like them better as people?

CHAPTER 13
MY CONCLUSIONS: THE SKEPTICS OF SCEPTICS CHALLENGE
TOPICS FOR DEBATE

There are two possible spellings for sceptic/skeptic. I regard myself as the first, given the side of the Atlantic I was born on. So I am sceptical of the orthodox stance on climate change. I conclude that humans have some part to play in climate change, but it is lost in the mix of natural climate mechanisms, and is very unlikely to be a huge or impending problem. I therefore define a skeptic as someone who is sceptical of sceptics (usually wholly dismissive of them) who rejects my premise and concludes that human emissions play a predominant role in climate change, or even the only significant role. I challenge the skeptics to review this book, and to be skeptical about my scepticism, rather than just to accept the status quo. If they do not question all premises, they are shunning the scientific principle on which enlightenment rests.

I am not interested in a he-said-she-said or yes-it-is-no-it-isn't ping pong pantomime. Only verifiable data and logical argument will do. In normal circumstances, one study is not enough to change direction, because the poor quality control of peer review is as likely to throw up a false outcome as a true one. A scientist should not be discounted because he got some research wrong in the past, or because he used to work as a geologist for a fossil fuel company. It is the data that matters, not the data collector.

I do not prescribe to a higher power, terrestrial or celestial, that will deliver us from evil. We have only science on our side to help solve our problems, so my faith is in it. I will be delighted to be proved wrong, as all those who value the scientific method should. Please try your best to knock down my rationale with any credible contradictory evidence you can muster. Launch your missiles against my edifice so that I can either reinforce its weak points or abandon it for a safer haven. I would be happy if I have got 80% of the 'facts' I mentioned in this book correct. That would be a lot better than the IPCC's record, if we strip away its mealy mouthed 'likelihood' and 'probability' assessments, and forced it to be straight with us.

My conclusions are currently as follows:

1. **There has been nothing unprecedented or particularly unusual about the climate over the last half century; the signs of AGW are not obvious.**

    a) The climate is about 150 years into a millennial warming phase which, on past experience (Bronze Age, Roman and Medieval Warm Periods), might last another 250 years or so.

    b) The Atlantic Multidecadel Oscillation was in a warming phase from about 1980 to 2007, and was probably the primary cause of the Arctic sea ice decline. On past experience, with ice peaking in the 1920s and around 1980, and at a low point in the 1950s and around 2007, the AMO should be in a perceptible cooling phase, and ice should be increasing, by 2030 (but who can know for sure?).

    c) Cyclones have not clearly been increasing in frequency or intensity over the last 50 years, the area of the world in drought is not increasing, the Leaf Index is increasing, and wildfires are reducing in extent overall, so the main fears of climate alarmists are without substance.

d) We can only say that there is 'climate chaos' if we recognise that the many natural mechanisms of climate change operate in chaotic ways that are impossible to predict with any degree of certainty. There is a 'climate emergency' but it is one of irrational panic in overreacting to perceived problems caused by the inevitable pursuit of human progress.

2. **It is reasonable to assume that human emissions have some effect on global temperatures, but we do not know how much.**

   a) If we accept the theory that doubling atmospheric CO_2 would, by itself, lead to an average global temperature rise of about 1.1°C, the Charney Range of +1.5°C to +4.5°C relies on positive feedbacks from the rest of the environment.

   b) Since we have a poor understanding of natural climate mechanisms, like the part that oceans and clouds play in moderating heating or cooling, negative feedbacks are perhaps just as likely. We just don't know how much positive and negative feedback there is, but there is a good chance that they roughly balance each other out, resulting in a transient climate response of 1.0°C to 1.5°C.

   c) This, and the fact that climate systems operate in a chaotic manner, explains why climate computer models get it wrong most of the time.

   d) If we were to discount the possibility of natural warming (or cooling) over the last half century, the anthropogenic warming during this period would be at the lower end of the Charney Range.

e) If the transient climate response is, say, between 1°C and 2°C, we should not be too concerned in the short to medium term. If we are then still worried about global warming, there is plenty of time for new technologies to be born and implemented.

f) On the basis that human population, energy use and therefore emissions level off as societies become wealthy, and then emissions start to drop as the high emissions biofuels (mainly wood) and coal are substituted for lower emissions gas and nuclear, we will probably not reach an atmospheric CO_2 level as high as 600 ppm.

g) The fact that there is not a good correlation between the rates of rise in human CO_2 emissions and atmospheric CO_2 confirms that our understanding of environmental systems and climate has some way to go.

3. **We cannot be sure whether some warming and more CO_2 in the atmosphere is good or bad for humans and the rest of nature.**

   a) Humanity seems to have faired better in warmer periods in the past. People in northern Europe and the north of the USA tend to migrate to, buy holiday homes in and have vacations in, warmer zones. We like the heat better than the cold, and many more of us die from the latter.

   b) The increasing world Leaf Index seems to suggest that some warming and more life-enhancing CO_2 are good for the environment.

   c) We have not studied the phenomenon of

changing ocean alkalinity long enough to draw any conclusions about the effects on sea life.

d) More CO_2 in the oceans should be beneficial to phytoplankton and seaweed but again there has not been enough research done to be conclusive, this time probably because this would be a good news story, and research money tends to be directed to perceived problems.

4. **Even if we could demonstrate that it would be good to reduce human emissions by a large amount, we currently do not have the technology to do so.**

   a) Wind and solar have a very limited effect on reducing CO_2 emissions because they are unpredictable and intermittent, and they rely on fossil fuel backup. They also use a huge amount of fossil fuel energy in their manufacture, installation, maintenance, related infrastructure and decommissioning.

   b) We are nowhere near inventing a method of long-term, grid-scale energy storage to make renewables viable.

   c) We have some way to go to make nuclear energy affordable.

   d) Biofuels are generally very harmful to the environment, except where they come from waste products, and then they can only play a tiny part in energy supply.

   e) The energy intensity required by cars, lorries and planes make petroleum products the only viable fuels. Car batteries rely on rare, expensive minerals (with environmental harm from large-scale mining), and on fossil fuel electricity generation for manufacture, charging and recycling.

f) What we do have just now are plentiful fossil fuels to help us become increasingly safer from extreme weather events.

g) Where we do cost benefit analyses on net-zero emissions targets, we find that vast (unaffordable?) amounts would be spent to achieve tiny (unmeasurable?) global temperature reductions, even if we assume that the technologies became available to deliver the objectives, and that the UN assumptions on emissions sensitivity to the climate are valid.

5. **There is a crisis of confidence among many intellectuals about human progress, where their utopian mindset believes that good is not good enough, and can even be bad.**

a) The very things which have given Western countries their prosperity, enhanced public services, well-being, freedoms and democracy are demonised in green ideology: capitalism, consumerism, international trade, economic growth, big business, fossil fuels, nuclear energy, more human brain power (more humans), plastics, GM technology, intensive agriculture, manufactured fertilisers, pesticides and herbicides, etc.

b) Gross hypocrisy abounds where public sector intellectuals and prominent figures in the arts bemoan the capitalist and consumerist wealth generation that affords them their livelihoods.

c) This hypocrisy becomes dangerous when neo-imperialist attempts are made to impose an impractical world view on less fortunate societies. A no-growth ideology means no progress for those who need progress most.

d) The situation becomes absurd when a carbon-based creature believes that it is beneficial or even possible to live in a zero-carbon civilisation.

e) The construction of a morally superior stance in a comfortable society demands the concoction of noble causes. Perspective and pragmatism are abandoned, and pessimism abounds. The activist cannot be proactive against imaginary demons, so the likes of the moaningly sanctimonious XR merely make a nuisance of themselves, like spoiled children.

6. **There is a crisis of confidence in the academic interpretation of science when political and ideological influences blur objectivity.**

   a) The scientific method relies on verifiable data to interpret how things are; too much of climate 'science' revolves around speculation of how things might be, in modelled or laboratory alternative versions of the real world.

   b) Since the mid-1990s climate science has followed political ideology, rather than the other way round. Questioning this ideology, and the scientific rationale which supports it, is likely to lead to being ostracised by academic authorities.

   c) Obtaining funding to investigate alleged

problems is much easier than to seek to prove that a problem does not exist.

d) Peer review is a poor form of quality control, and it is difficult to obtain funding for replication studies to verify or disprove existing research.

e) The alleged problems caused by anthropogenic climate change will not be solved by speculative, academic modelling and ideological consensus, but by irreverent, out-of-the-box entrepreneurial risk-taking that finds technological solutions that no one had previously thought about, or believed would work.

7. **The economies of the democratic countries which apply inappropriate renewables technologies are being harmed, and those in autocracies like China, Russia and Saudi Arabia are benefitting.**

a) High energy costs, due to the widespread adoption of renewables, affect all areas of the economy, with international industrial competitiveness harmed. Liberal, democratic countries are scoring an own goal.

b) High energy costs also erode the personal standard of living, and exacerbate fuel poverty in wealthy countries.

c) While it could be argued that exporting industry to a democracy like India is beneficial, the same cannot be said for aiding an autocracy like China; and there are more efficient ways of helping developing countries like removing trade barriers and funding basic infrastructure.

d) By pretending that renewables significantly reduce our reliance on fossil fuels, and acquiescing with green demands to stop extracting domestic fossil fuels, we merely import them from nasty regimes like those in Russia and Saudi Arabia.

e) Importing fossil fuels instead of extracting them from plentiful domestic reserves, for example by fracking in Britain and Germany, also increases CO_2 emissions from transportation.

f) By trying to stop poorer countries using cheap fossil fuels to fight their way out of poverty, Western democracies are guilty of cruel neo-imperialism, on which Russia and China capitalise.

8. **Human development sometimes damages the environment, but the richer we get the more our resources can be used to repair and sustain it.**

   a) The climate will change because it always has and always will, even if human activities now make a contribution. Homo sapiens and other species will do what they have done before: adapt their lifestyles, migrate to more sympathetic climatic zones or become extinct. We have shown that we are masters of adaptation.

   b) Wealthier people look after their environment better and apply conservation measures.

   c) Other species now have something they never had before to adapt to climate change: a caring humanity which enjoys seeing them around, is monitoring their health and is taking positive steps to help them survive.

9. **There are huge forces at large to maintain the status quo which are commercial, political and philosophical, and which are at odds with scientific reality.**

   a) A huge number of businesses, research units and universities currently benefit from the eco-industrial complex, and will resist questioning the veracity of the present climate change orthodoxy. Profits are essential; integrity and veracity seem to take second place.

   b) Democracy does not require integrity of politicians; in fact it is a handicap if it stops them bending to public opinion or being seen as insensitive to minority interests. Democracy does not ask politicians to do what is best for the electorate, but to apply measures which they reckon the electorate perceives will be good for it.

   c) A huge number of academics, politicians and media hacks have staked their reputations on promoting an unrealistically gloomy future because of climate change, and continue to spread fear of possible future scenarios to keep the issue in the spotlight.

   d) The scientific method requires people to seek out and accept dry facts, if necessary welcoming the realisation that their previous world view was wrong. This is psychologically difficult, and the threat of contradiction makes them lash out irrationally at their critics and call them 'deniers' of their dogma.

   e) Emotions often trump logic when we have enough privilege to be contrary, rather than be disciplined by the pragmatism required to achieve more basic needs. This is best illustrated by the adoration of the rantings of a troubled teenager with serious mental health problems and little knowledge of her alleged specialist subject, climate change.

10. **When reality and common sense prevail, it is possible that a more pragmatic energy policy can be adopted which will be good for both developed and developing nations, particularly democracies.**

   a) Survey after survey shows that climate change comes way down the list of issues that the general public thinks is important, which will only be reinforced as energy cost and security problems escalate. Politicans are reacting to vocal pressure groups rather than the silent majority.

   b) Many more businesses are handicapped by the high energy prices caused by current policies than benefit from publicly sponsored renewables subsidies.

   c) If enough of the Western media question alarmist stories and predictions with scientific rigour, instead of just accepting them as cheap copy, there could well be a tipping point for public opinion, with politicians turning like a weather vane.

   d) This tipping point could result from escalating energy prices, power cuts due to over-reliance on renewables (unreliables) and insufficient gas and nuclear power (reliables), or by just being over-pressurised to give up the freedoms afforded by, for example, affordable petrol-fueled cars, gas heating boilers or flights for business and holidays. The situation in Ukraine should make us question everything to do with energy supply and security.

e) The tipping point for showing that the science is flawed might be an observable upwards trend in Arctic sea ice by about 2030 (but we can't rely on it because the only certainty we have of the climate is that it is uncertain).

f) The damage done by inappropriate energy policy could be mended within one or two decades (by about 2035) as wind and solar projects reach the end of their useful (useless?) lives (assuming there is not a breakthrough in mass energy storage by then).

g) If we abandoned net-zero, the increased economic power would allow democracies to exert more influence on autocracies, rather than kow-tow to them to obtain essential fossil fuel energy and cheap manufactured products, as well as help other countries out of poverty and protect the environment better.

CHAPTER 14
ADDRESSING CLIMATE CHANGE, THE STORM IN A TEACUP
A LESSON IN PRAGMATISM

Every year from 1995 until 2019 people interested in climate change would meet at some point in the world at a COP event with the intention of solving the problem, once and for all. Before the event, academics would vie to present eyecatching research to highlight the impending dangers of failing to take human-induced climate change seriously enough. The environmental correspondents of newspapers would lap this up, promoting the alarming messages without question. Participants would gather in a mood of collegiate fervour, with famous actors, pop stars and the climate glitterati adding celebrity, all convinced that this time unchallengable sense would prevail. But it never did. We have had a quarter century of absolute failure, and Glasgow 2021 was exactly the same.

After the initial optimism, it would become clear that perhaps there was not consensus after all about how to move forward. Drafting clerks would explore every word of every option with political representatives from all the countries until, at the very last moment, a resolution would be agreed, invariably so loose that representatives would return home satisfied that they could claim the best of green motives but be committed to very little. They had been banging their heads against a brick wall, but they never seemed to learn the lesson that this was inevitable, and always will be if the current approach continues. The wall they come up against is present because no one knows what strategy/technology will actually reduce emissions significantly and, even

Our last chance to solve the climate crisis (cartoonsbyjosh.com).

if there were a ready-made solution, no one wishes to report home that their citizens will need to suffer financially, and give up lots of their comforts, to deliver it.

Despite all the bluster, my contention that climate change is 'a storm in a teacup' can be justified, at least in comparison to various other problems that the world has to face. The world has warmed up by just over 1°C since around 1800 when we were still in the Little Ice Age. Whatever the mix of natural and human-made causes, that is surely a good thing, because warmer winters save lives. The previous warm periods were probably about 2°C above the LIA baseline (a temperature that, on the basis of current knowledge, will probably only be reached during the second half of this century), so we know that humans and other species survived and even prospered in such conditions. More CO_2 in the atmosphere is unprecedented in hundreds of thousands of years (actually 2.6 million years, but so what?), but it is good for plant life and consequently animals like ourselves.

Even if we were to experience more intense storms with the likely temperature rise, we are good at adapting, not just in rich countries but also in poorer ones. On 12 November 1970, a Category 3 storm with a six-metre tidal surge hit East Pakistan (now Bangaldesh) and killed more than 250,000 people. Gradually a programme of public awareness, early warning, storm shelters and emergency response systems was developed, such that the number of people dying due to intense storms by the 1990s was about 15,000 per year, until the 2010s when the figure dropped to an average of 12. Lomborg: "For more than half of the population of coastal flood plains throughout the world, each dollar spent on protection will avoid more than $100 of damage."

There may be some areas subject to more flash floods, especially a country like Bangladesh with the power of the Ganges and Brahmaputra to contend with. Lomborg: "Sixty per cent of Bangladesh is vulnerable to flooding, and 67% of the Netherlands is similarly vulnerable .. The Netherlands suffered a devastating flood in 1953, the Watersnoodramp. More than 1,800 people died when water breached the dykes… Since 1953, there have been just three floods and one fatality to flooding… In the first two decades of the 21st century, more than 3,000 people have died from flooding [in Bangaldesh]… As Bangladesh gets richer, it will be able to spend more on adaptation… While the initial cost is about one per cent of Bangladesh's GDP today, it will be one-tenth or less in the 2050s." This is based on UN projections of economic growth in developing countries.

Higher temperatures might make vegetation more combustible, but we have also demonstrated that we have this in hand because the global extent of wildfires is reducing. There may be some areas subject to more droughts, but again we are good at irrigation schemes. As long ago as 1904 the British completed the first Aswan Dam to regulate the waters of the Nile in Egypt, either to control floods some years or to enhance flow in drier years. The $4.8 billion Grand Ethiopian Renaissance Dam will provide 6.45 gigawatts of electricity as well as irrigation when the ten-year period for filling it with water finishes around 2030. The cost was about 5% of Ethiopia's GDP in 2017, but the supply of plentiful electricity will rapidly expand the wealth of the region. The lake formed will also allow the harvesting of 7,000 tonnes of fish each year and will be a tourist attraction to bring in foreign currency.

We have proved that increased wealth and technological progress are a match for everything the climate can throw at us, which is why I placed economic growth in developing countries as the first priority. I would also wish to decouple climate change from biodiversity loss, since it is generally agreed that the former currently plays a minor role in the latter, and we know how to solve the other human factors harming the environment. So for the foreseeable future, I would prioritise economic development in developing countries first, protection of wild habitats and species second and anthropogenic climate change third. That does not mean that we do not address them all, but we should not adopt policies for a lower priority cause that significantly harms a higher cause. Unless we get people out of poverty, they will be unable to contribute to environmental improvements and initiatives to lower human emissions (if we definitively conclude that the last objective is possible and necessary). Imposing expensive energy policies will be counterproductive. Cutting down forests, or apportioning large tracks of land for energy crops to produce biofuels, also contradicts my category prioritisation.

IN FOR A PENNY, IN FOR A POUND

In Chapter 10 I referred to Bjorn Lomborg's work to assess the value for money of climate change initiatives. He assembled 50 teams of economists, with several Nobel laureates among them, under the auspices of the Copenhagen Consensus thinktank.

Using the median IPCC model for how the climate might change for the rest of the century (remember, observations support the bottom end of the possible range, not the median), they made assumptions to work out the social, economic and environmental benefits from investment in various global problems for every dollar spent. In his book he does not say whether he addressed all the 169 targets in the UN's Sustainable Development Goals, but he quoted 14 study areas to do with trade, gender, health, food security and nutrition, climate and energy. Doubling renewable energy and applying measures aimed at keeping the average world temperature less than 2°C above pre-industrial levels (assuming they worked) provided less than a dollar return on each dollar invested. This is unsurprising given how inefficient renewables are. It makes no sense to invest in something with negative financial returns, but that is what current energy policy does in many countries.

As we saw in Chapter 10, when he gauged the long-term investment against net present value, the investment gave a strongly negative figure. The best way for a non-accountant to understand this is to imagine our much poorer grandparents investing in something which will only give a return in seventy years' time, like our current approach to climate change. The value of the reward will inevitably be tiny compared to our current wealth. To obtain value for money we need to obtain shorter-term benefits, certainly less than 30 years. In the dozens of public private partnership school and hospital projects I was involved in during the 2000s, many 25-year funding packages (by the private sector) with lease back (to the public sector) were based on the annual charge paying off the capital over about 20 years. The last five years' payments provided the profit to contribute to the pension pots of the people whose combined investment in pension funds paid for the new buildings in the first place, and who were then close to retirement. The 25-year period was about the longest viable for such investments because of the limits of risk management projection and the single-generational period of return on investment.

Lomborg used William Nordhaus' DICE model to review five scenarios for addressing climate change, from doing nothing (zero investment) and assuming an average temperature rise of 7.4°F (4°C) by the year 2100, to the biggest spend to restrict the temperature rise to 3.9°F (2°C). Probably no one believes that we can/will adopt measures that might be needed under a 1.5°C rise scenario, so he did not consider that. He combined in each scenario the cost of climate change (adaptation) with the cost of implementing mitigation measures to determine best value for money. The optimal solution was the second scenario, where we accepted the cost of a 6.3°F (3.5°C) rise in temperatures ($87 trillion) with a relatively small investment in climate policy initiatives ($21 trillion). Restricting the temperature to 3.9°F (2°C) cost about twice as much. There are many uncertainties and imponderables in any long-term predictions, but the best Lomborg could do was adopt the same parameters as the UN and draw pragmatic, logical conclusions. In other words, we should major on our well-tested, value-for-money adaptation measures and spend less on inefficient and unproven alternative technologies.

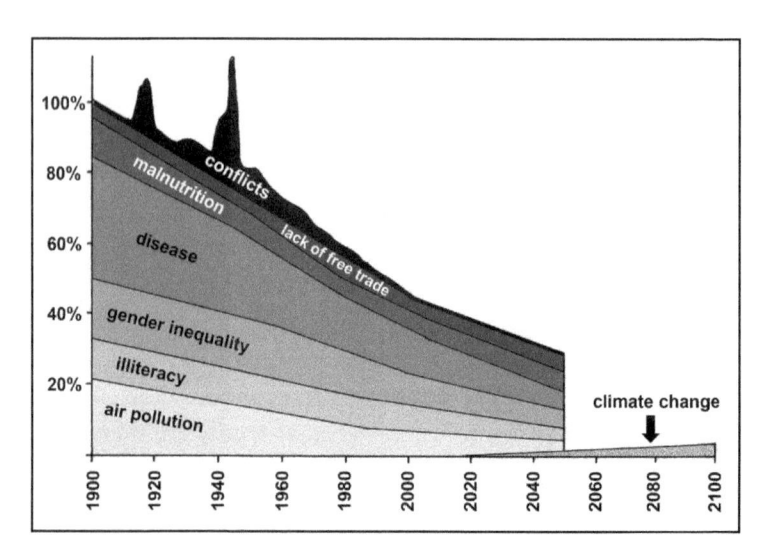

Figure 40 Cost of world problems as a percentage of global GDP.

The Copenhagen Consensus considered how much richer the world would have been if we had solved various issues between 1900 and 2050; in other words, what was the burden in terms of percentage global GDP of each problem? This is shown in Figure 40. Again, this is an illustration of how climate change, whose cost will only be a few percentage points of global GDP at worst by the year 2100, really is a storm in a teacup compared to the other problems which have burdened us in the last 120 years. We should continue the progress in areas we have the formula to resolve rather than burdening ourselves with climate mitigation measures whose effectiveness we cannot easily predict. The downward trend in the century from 1900 has moderated after the year 2000, perhaps suggesting that we have taken our foot off the pedal on the problems which matter most. There is still much to do for the poorest billion, never mind the aspiring 3 billion above that, who are defined as Level 2 on Hans Rosling's scale, before we can expect them to start adopting expensive alternative energy technologies.

FOCUSING ON WEALTH GENERATION FOR POORER COUNTRIES

The best value investment in the Copenhagen Consensus exercise was implementation of free trade measures under the World Trade Organisation's Doha Development Agenda. "If we were to successfully conclude such a treaty on freer trade, economists estimate that it could by 2030 make the average person in the poor world a thousand dollars richer per person per year." A return of $2,011 on every dollar spent is probably so large because little investment would be needed, just the good sense of richer countries to stop protectionism and allow poorer countries to trade on a level playing field. The EU agricultural policy is one of the biggest offenders, with hundreds of thousands of farmers being compensated for their inefficiency. The policy was originally intended to provide security of food supply in the decades after the Second World War. It seems reasonable for governments to cushion the shock of social and economic change with short-term subsidies, but not to perpetuate intergenerational inefficiencies which have an unfair knock-on effect on poorer countries.

Adam Smith said that we should use the money saved by cheaper imports to invest in new technologies for our future benefit, but with this sage advice it still took many decades for Britain to repeal the Corn Laws in 1846 due to the vested interests of landowners, while industrialists and urban dwellers wanted cheap food from America and elsewhere. If EU citizens paid less for their food, they could invest it in other life-enhancing activities. They might even invest it in shares of companies in poorer countries, should their growth potential be converted into dividends. Rather than subsidies and handouts, which are often squandered, investment in self-help, basic infrastructure and commerce benefits everyone.

Instead of environmentalists marching in protest to EU governments on the basis of the imagined effects of warmer temperatures on poorer nations, it would be much more pertinent to march against cruel trade protectionism, and the use of biofuels which the World Bank (after *Ivan, Martin and Zaman* 2011) reckoned had raised global food prices by 75%. "*Guardian* columnist and strong climate campaigner George Monbiot called the subsidies driving the biofuel industry's growth 'a crime against humanity'." The increased wealth of poorer countries due to fair trade would help them to overcome age-old human problems (*Tol and Dowlatabadi* 2001 calculated that an average of about $3,000 GDP per person per year was the level at which malaria is conquered), take measures to protect themselves against extreme weather events (whatever the cause) and exceed a threshold of about $5,000 GDP per person per year when people stop obsessing about survival and start protecting their environment.

According to the International Monetary Fund in 2019, the world average GDP/person/year was $11,355, but the only African countries above $5,000 are Equatorial Guinea ($8,927), Gabon ($8,112), Botswana ($7,859), South Africa ($6,100), Namibia ($5,842) and Libya ($5,019). Algeria, Morocco and Tunisia just scrape into the $3,000+ category, but even countries like Egypt, Angola, Ghana, Nigeria and Kenya fail to make it to this level. India ($2,171), Bangladesh ($1,905) and Pakistan ($1,388) have some way to go to make up the gap between them and China ($10,098), which itself is unlikely to move away from its reliance on cheap coal for many decades.

We might expect certain Central and South American countries to care more for their environment, for example Uruguay ($17,029), Chile ($15,399), Mexico ($10,118), Argentina ($9,887) and Brazil ($8,796). But it is understandable that a country like Indonesia ($4,163) will seek to turn more of its land into wealth-generating food and palm oil production. And how can countries with important natural environments such as Mozambique ($484), Madagascar ($463) and Malawi ($370) contemplate more than basic subsistence? Countries like the USA ($65,111), Sweden ($51,241), Germany ($46,563), France ($41,760) and the UK ($41,030) have no right to lecture poorer countries on their energy use and environmental credentials when economic growth must be their fundamental priority.

Lomborg: "Avoiding malnutrition in the first two years of a child's life costs about $100. Because good nutrition helps develop the child's brain, it leads to better educational outcomes and phenomenally higher productivity in adulthood... Every dollar spent on fighting early childhood malnutrition results in $45 of social good... Tuberculosis mostly kills adults in their prime, leaving children without parents. For about $6 billion annually, we could save nearly 1.6 million people from dying each and every year... Each dollar spent on contraception and family planning education will generate $120 of social benefits across the most vulnerable societies." In other words, a certain amount of investment pump priming – gets fast and effective results if applied properly Where else would we get a 45-fold or 120-fold return on investment? When you see dramatic pictures of an extreme weather event on TV in a rich country, tragic though their consequences can be for the dozens or hundreds of people badly affected, remember that 4,400 people die each and every day from TB in poor countries, despite this being an easily treated condition.

While Hans Rosling calculated that the world's population would level off at about 11 billion, others have estimated that a programme of economic development among the poorest nations would reduce that to about 9.4 billion, on the basis that average birth rates would rapidly drop below the replacement level of 2.1. If Bangladesh (GDP/person/year $1,905) and Nepal ($1,047) can achieve this level, and Thailand ($7,791) and Azerbaijan ($4,698) can have birth rates of 1.5 and 1.7 respectively, merely getting the poorest countries up to more than $2,000 GPD/person/year would make an enormous difference. It's not that we hate the rest of humanity so much that we want less people, but the less human beings we have, the less resources of all types we will need, including the only trustworthy mass energy source for the foreseeable future – fossil fuels.

I said above that we should decouple climate change and protection of the environment – how can the Royal Society for the Protection of Birds in the UK wholeheartedly campaign against migrating seabirds being chopped up by offshore wind farms when it is a member of Climate Coalition? We should do the same with the fight against poverty, which must be the top priority, even if it inevitably means a short to medium-term rise in emissions. During that time, climate change will not have a big effect on poor communities if their increased wealth allows them, like Bangladesh, to address extreme weather events. Unfortunately, many international aid agencies have been contaminated by neo-Marxist dogma, with the focus on political ideology rather than pragmatism.

Oxfam's 2020 report with the Stockholm Environment Institute, *The Carbon Inequality Era*, draws attention to the wealth and emissions gaps between countries. This seems to me irrelevant and distracting politicisation by Tim Gore, a co-author, and Oxfam's Head of Policy, Advocacy and Research (previously with Climate Action Network Europe and the left-wing thinktank the Fabian Society), and Oxfam Chief Executive Danny Sriskandarajah (previously Deputy Director of the left-wing Institute for Public Policy Research). The latter complained in the accompanying press release: "The over-consumption of a wealthy minority is fuelling the climate crisis and putting the planet in peril." The report demonised the extravagant richest 10% but did not mention that emissions in rich countries, particularly America, have levelled off in recent decades and started to reduce, mainly due to the transfer from coal to gas, or that economic development of poorer countries reduces the gap between them and richer ones. In developed countries, the gap between the richest and the poorest (comparatively poor) is increasing but that should not be of relevance to Oxfam, unless their objective is to make rich people feel guilty for their wealth so that they donate more to charities.

DEVELOPMENT OF DEMOCRACY AS AN INTEGRAL PART OF ECONOMIC DEVELOPMENT

Zimbabwean Geoff Hill, Africa correspondent for *The Washington Times*, in an essay for the Global Warming Policy Foundation called *Heart of Darkness: Why Electricity for Africa is a Security Issue*, pointed out that the lack of electricity from the grid in African countries prevents investment in jobs, leading to unemployed young men who turn to crime and militias. He also noted that billions of dollars had been supplied in aid over many decades but that corrupt, undemocratic governments had diverted most of it away from the intended uses. Although democratic institutions take a lot of money to maintain, the path to full democracy must go hand-in-hand with economic development so that the wealth is spread, which in turn leads to more growth. He made comparisons between North and South Korea, East and West Germany, Haiti and the Dominican Republic, Botswana (a democracy with two thirds of its people with electricity) and Angola (a non-democracy with oil riches but only 41% of its people have electricity), Kenya (a democracy with almost full-citizen access to electricity) and non-democratic Uganda (26%) and Tanzania (33%).

Hill quoted Harare engineer Aaron Chiwoko: "Angola and Nigeria are among the world's top-ten oil producers. Mozambique has natural gas. South Africa, Botswana, Zimbabwe, Kenya and others have billions of tons of coal in the ground. But they can't use any of these at home with funds from the World Bank or even with loans from some democracies who decree there must be no fossil fuel in the mix. And so electricity is either not available, or it's in short supply with long outages. And even those who have it only use the grid for lights and maybe a fridge and TV because anything that generates heat uses a lot of power. For that they turn to paraffin or, more often, firewood." People often forget how cold it gets in winter at night in many African countries, which has ten of its capital cities at elevations between 1,250m and 2,300m above sea level, but only two of these have more than 60% of their citizens with access to electricity. Forty of the top fifty fastest-growing cities are in Africa as people move from the countryside looking for work.

It is vital that rich donor countries have assurances about good governance and proof that humanitarian development funds are being used for their intended purposes, but they should not dictate the nature of a country's development plan, which will inevitably seek the cheapest energy source available, until it is wealthy enough to choose others. Because of the West's impractical high-mindedness, China is supplying aid and coal power technology. When it demonstrates that an authoritarian regime gets results, democracy takes a hit.

A PRAGMATIC ENVIRONMENTAL POLICY

We might expect the WWF *Living Planet Report 2020* to come up with solutions to the problems facing us, but it majors on spreading alarm. A cynic would accuse the academic contributors of generating this to attract further research funding, although I am sure they believe their own hype. As mentioned in Chapter 6 the WWF has constant references to climate change. The word 'climate' occurs 144 times in the 107 pages of text, whereas 'poverty' only appears five times and 'reforestation' only twice. Robert Watson of the Tyndall Centre for Climate Change Research had the privilege of writing the first chapter. "Climate change **projections** show that food production shocks across sectors are increasing, and are **likely** to worsen as extreme events such as marine heatwaves and drought become the '**new**

normal'." This is all speculation; futurology and not science. The report also mentions a Global Risks **Perception** Study with extreme weather and climate action failure as the top risks. 'Perception' is the opposite of scientific objectivity.

As I showed in Chapter 5, the contention that the Great Barrier Reef is being destroyed by climate change is a modern myth, but the WWF report perpetuates this: "Ocean heatwaves have already destroyed half of the shallow-water corals on Australia's Great Barrier Reef. [no they haven't!] As the IPCC reported in 2018, scientists have projected that a 2°C global temperature rise will result in the almost complete eradication (a 99% loss) of coral reefs globally." So were the reefs eradicated during the Bronze Age, Roman and Medieval Warm Periods?

Mathis Wackernagel, President of the Global Footprint Network, is a prominent contributor. He reckons that we are exceeding the planet's capacity to sustain it and us by 56%, but he has to allocate 60% of the problem to the inability of the planet to absorb the CO_2 emissions we are producing, otherwise his graph of Humanity's Ecological Footprint would not look half as scary. His assessment no doubt relies on the CO_2 residency time of the UN Bern Model (Figure 22 in Chapter 8), whose validity I have questioned. If the half-life of CO_2 absoption into the biosphere is around ten years instead of 29, and the 'bath' is not filling up as quickly as assumed, Wackernagel's bubble would burst. The land taken up by cities and agriculture, and the encroachment on forests for timber and oceans for fishing, has risen from 4 billion hectares in 1961 to 8 billion in 2020 (a 100% rise), an inevitable result of the global population rising from 3 billion to 7.8 billion (a 160% rise). Without CO_2 emissions, the sum of these other human impacts fall well below his dotted green line of Earth biocapacity. Once again, an academic is playing around with climate and ecology models with results from the fantastic to pure fantasy.

Instead of illustrating success stories and demonstrating how these can be continued by pragmatic action (as I did in Chapter 6), the report exudes a high-level, high-brow ideology with "human rights and moral philosophy . . [which] provides a starting point for hopeful conversations". It places its faith on computer modelling. "Our **imagination** creates the new worlds we could live in. Building **digital twins of our living planet** allows us to better imagine many possible future states of life on planet Earth. The **in-silico twins of the Earth** are created in computer models of the Earth system and its diversity." (It does not consider strategies for invasive species because of "the lack of models and scenarios of biological invasions". So, no computer model leads to incomprehension and no policy?) In other words, if we manipulate in computer models the hopelessly inadequate data within the impossibly complex ecosytems, we can produce whatever imaginary worlds we desire. The real world does not work like that. Let's stick to what we know has worked in the past and accelerate this.

We can work with nature. "Diversified agricultural landscapes can support much more biodiversity than was previously thought. Agricultural systems can also be managed in such a way that they facilitate, rather than constrain, species dispersal through corridors and along migratory routes." Of course they can, and they do in rich countries where examples are many. We need to continue this, not fill our fields with wildlife-starving monoculture biocrops. The report is notable for failing to condemn biofuels. In one place it suggests GM technology as part of the armoury to intensify agriculture, but in the next paragraph mentions organic farming which does the opposite, no doubt to tick a box in the minds of its green constituents.

We can set aside land for nature. To date we have allocated 15% of the world's surface to nature reserves, although our husbandry of at least a third of this needs to improve. WWF: "About 25% of terrestrial Earth can be considered 'wilderness'… [but] most of this is contained in just a small number of nations – Russia, Canada, Brazil and Australia." So, if we could retain the wilderness areas, and increase protected areas to 25%, with all ecosystems represented, we would obtain the 50% target that WWF hopes for, and we could do this within decades, given the progress made so far. A similar strategy is taking place with the oceans, and the WWF report barely mentions the opportunity for onshore fish farms to replace wild sea fishing.

WWF asks us to waste less food, which is only common sense, but the gains are likely to be tiny in overall terms. In growing vegetables in my garden I am constantly amazed at how inefficient plants are, because they produce a huge amount of plant material to collect the sun and the earth's nutrients for the small part we can harvest. Many times more plant waste ends up on my compost heap than I consume. GM technology holds the key to improving this. I love vegetables but I also enjoy meat, so the report's prescription for a modern diet will be resisted by most people: "whole grains, fruits, nuts, vegetables, beans and pulses; and fish, eggs, dairy and other animal-sourced meats in moderation where desired". Burgers will continue to be desired, as witnessed by the 39,000 McDonald's restaurants in 119 markets. William van Wijngaarden's calculation, mentioned in Chapter 8, was that increased methane from belching cows and other human factors will only raise the temperature by about 0.1°C in the next century.

It is difficult to make a purely economic case for environmental projects but, from the comfort of our human-made environs, we seem to instinctively value the birds, bees and natural beauty around us. In our emotional response, we should be careful not to fall into ideological or political traps that focus our minds away from the proven positive responses we can make. Remember that globally we have turned the corner on deforestation, we are increasing the area of protected land and sea, we are better defining the areas of the planet and its wildlife that are in danger and we are better at eradicating invasive species on islands. Instead of wallowing in the gloom, as environmental (in)activists do, we should reinforce the many good things we are doing.

A PRAGMATIC ENERGY POLICY

My contention is that, especially in the wealthiest countries, we should spend less money on energy initiatives which seek to address anthropogenic climate change, but which work inefficiently and provide poor value for money, so we can focus resources on strategies we know work well to combat world poverty and protect the environment.

1. **We must phase out all subsidies for biofuels, unless it can be demonstrated that the fuel comes from waste products.**
   The reasons for this should be obvious by now. This would include the banning of E10 petrol in the UK, which contains 10% bioethanol, whose production may cause environmental harm.

2. **We must phase out all subsidies for solar and wind, unless they contain a facility to store energy that evens out the fluctuations imposed on the grid network.**
   The national grid should not be left with the consequences of the failings of solar and wind. Domestic solar panels should be able to store the energy needed by a house to meet the peak evening demand after the sun has set. A similar condition should apply to solar farms. Wind farms should store energy, perhaps for several days, when the presence of a high pressure region means windless conditions. If this means a sharp fall in the construction of such renewables because domestic day-scale storage and regional multi-day, grid-scale storage are not viable, then so be it. Intermittent wind might be used to manufacture hydrogen for use in transportation if diesel fuels cannot become less polluting, or if the economic case for this as a means of reducing carbon dioxide emissions can be made.

3. **That means we could greatly reduce subsidies on fossil fuels, which may always be necessary to some extent to meet peak demand, but should not be used predominantly to make up for the**

intermittency of renewables. **Penal taxes on fossil fuels are also folly.**
It is crazy to subsidise fossil fuels purely to support the use of inefficient, already subsidised renewables. Fossil fuels are good tax-raising targets for governments because of their popularity, but penal taxation is not a practical way of reducing emissions. Only rich Western democracies would adopt high energy taxation that would weaken their economies, the poorest in these societies would suffer most, and there is no point in dissuading people from using one energy source if another one that effectively reduces emissions is not available. The IMF reckons that a carbon tax rate of $75 a tonne is required globally to put people off fossil fuels, while the UK's carbon price was about $24 a tonne in 2020. Tripling it would be economic suicide.

4. **Because gas and oil will be necessary for the foreseeable future for grid energy reliability and for materials and products made from petrochemicals, modern technology (fracking) should be used to obtain this as close to demand as possible to reduce imported emissions and supply costs. The same applies to coal extraction for industries that rely on it.**
This is merely common sense and good housekeeping. The UK must restore its 'dash for gas' policy to ensure reliable energy into the 2030s.

5. **The direct and indirect impact on the environment should be fully assessed, and any issues addressed, before any form of energy facility is allowed to proceed. Monitoring thereafter is essential.**
We should not have (overly) strict safety and environmental rules for fossil fuel extraction facilities like hydraulic fracturing, when we allow lax rules for

renewables. Perhaps we should retrospectively fit cameras and other detection devices to wind and solar farms to record how much wildlife – birds, bats and insects – are harmed. Some experiments have indicated that non-white wind turbines are seen better by birds. We must not ignore the environmental damage done to far-away places by the likes of lithium mining, or the disposal of toxic batteries at the end of their lives.

6. **A cost/benefit analysis must be made public for all energy strategies and developments, including the effect on emissions, again directly and indirectly. No embedded emissions or associated infrastructure costs should go under the radar.**
These are currently sadly lacking, and the public is being deceived. The government should be honest enough to admit that some aspects of an energy strategy rely on technologies which are not mature or proven, and the costs are therefore impossible to estimate with any accuracy.

7. **Energy policies should be appropriate for a country's stage in economic development.**
Rich countries should not expect poorer ones to adopt the same policies that they are implementing, unless they intend to pay for them, which they will not. The cost of attempts at net-zero are expensive enough for citizens of rich countries without asking them to triple or quadruple their energy bills to subsidise the rest of the world.

8. **Independent research must be carried out to determine more accurately the transient climate response and other uncertainties in climate science. A range of between 1.5°C and 4.5°C is not good enough to allow credible forward planning.**

We still do not know how much of the warming in the last 200 years has been natural and how much has been human-produced. Even the median of the Charney Range is probably much too high an assessment of the impact of emissions. The Bern Model must be interrogated to better understand why it is believed that CO_2 stays in the atmosphere for so long, to continue acting as a greenhouse gas. If the conclusion is that doubling CO_2 in the atmosphere only leads to a global increase of 2°C or less, we can all relax a bit. Instead of research predominantly defending the status quo, and sidelining unconventional people like Shaviv, Svensmark and Haigh, all avenues should be properly explored.

9. **Quality assurance protocols for publicly subsidised academic research must be introduced which are significantly better than current peer review.**
If, as alleged, academic research gets it wrong as often as it gets it right, we must take steps to obtain better value for money for the billions spent each year. Peer review is often effectively self-regulation. Independent government spot checks, and insistence on a proportion of spending on replication studies, would embarrass universities and scientific journals to get their own acts in order.

10. **In case it is definitively proved that increased atmospheric CO_2 is a large problem, rich countries must encourage research in new technologies, whether for energy generation and storage, or for carbon sequestration.**
It is inevitable that fossil fuels will be our predominant energy source until at least mid-century because no other viable and affordable

technologies are in sight. Emissions will increase until then as countries grow more wealthy, and we will have to respond to the consequences. Meantime, it is the role of rich countries to research low carbon energy technologies, just in case solid scientific evidence proves that their implementation would be beneficial. Much more funding must be directed towards this. Lomborg: "Globally, in 2020, taxpayers will pay $141 billion to subsidise inefficient solar and wind energy. This will buy us just $6 billion in actual R&D. Instead, we should spend $100 billion directly on research and development."

A new, affordable, emissions-light, mass energy source would allow us to reconsider electric transportation and hydrogen power. Nuclear may be the best bet, either large-scale fission or mass-produced, small-scale, unitised fusion. It is difficult to envisage grid scale energy storage to make wind and solar viable, but who knows? A hundred years ago we could not have imagined being able to extract large quantities of the invisible gas nitrogen from the atmosphere and turn it into a solid form for fertilisers to feed seven billion people, so perhaps we will do something similar with carbon dioxide. Or perhaps we will be able to capture the CO_2 from fossil fuel electricity generators at large scale and at a realistic cost, similar to the way we previously did at the Drax power station to filter out harmful pollutants from coal generation. We would then need to work out where to store vast quantities of this gas such that it will not escape back into the atmosphere, which it might do if we use the underground voids left over from old coal, oil and gas seams.

Or someone might come up with a completely different technology that we cannot possibly imagine just now. Or we might prove conclusively that more CO_2 in the atmosphere, on balance, is a thoroughly good thing.

CONCLUSIONS

As a humanist, I do not believe in a supernatural deity which controls the elements. I also do not believe that human beings are the devil's children whose greed is defiling the planet. I observe that their wealth and resourcefulness, brought about by employing intensive agriculture and exploiting fossil fuels, is increasingly being used to cherish and enhance the rest of nature, and repair damage done in reaching this stage in the fleeting time of humankind. Utopian concepts like 'saving the planet' and 'net-zero emissions' have no place in an enlightened society, and are a threat to it. We have good cause to feel hopeful for the future, given the progress we have made so far, but only if we reject the pseudo-religious rhetoric of activists and stick to sound science.

Despite years of searching I can find no 'scientific consensus' on climate change. Remarkably, most politicians in Western democracies believe that there is, so there is a 'political consensus'. But how can the science be settled when 40,000 academic studies on the subject are carried out each year? The climate picture is made up of a great many jigsaw pieces. Each academic reckons that more needs to be learned about their piece of the puzzle, otherwise research funding would not be forthcoming, but that the overall picture is pretty clear. Yet I have found that some pieces have been forced in and do not fit properly, many pieces are missing and some have been ignored because they do not accord with the generally perceived vision of what the overall picture should look like. But we do not have a picture on the front of the jigsaw box to help guide us. As democracies suffer increasing energy costs, large parts of the jigsaw need to be reviewed and rebuilt, but at the moment there is no authoritative, unbiased entity which can do this.

I have spoken to politicians who know that something is seriously wrong, but fear ridicule if they depart too radically from the status quo. So they might purchase an electric car despite it having been pointed out that the net emissions over its lifetime will not be much different to an efficient petrol vehicle. They are just spending a lot more on a virtue-signalling exercise to gain what they believe to be public credibility. They do not wish to be labelled climate change 'deniers', failing to appreciate that the use of this personal insult demonstrates the lack of rationale and knowledge on the part of the activist calling it out. But I know they will come on board if the tipping point of public opinion is reached when the cost of living gets to a certain level. Enough will be enough and the electorate will decide, voting in people as objectionable as Donald Trump to make the point if more reasonable politicians do not face the facts.

I have been a long-term supporter of the National Trust which is a custodian of natural and built UK heritage, and more recently a member of the John Muir Trust which seeks to protect wild areas. I would describe myself as a committed environmentalist, but the last thing I would like to be called is 'Green'. I have always associated the UK Green Party with an anti-capitalist, anti-progress agenda, and the politicisation of groups like Greenpeace, WWF and Friends of the Earth has been a turn-off. The unscientific, unenlightened ideology which has pervaded the so-called Green movement has poisoned the term 'Green'. Perhaps we need a new word to dissociate ourselves from this.

Let me suggest 'Blu' environmentalism. I go for the colour of the sky and the oceans rather than that of chlorophyll, a carbon-based compound which also contains hydrogen, oxygen, nitrogen and magnesium. I have chosen a phonetic spelling rather than the normal English word to highlight the separate meaning: 'having a concern for the environment that is pragmatic, based on scientific principles, and is not at odds with the progress of humankind'. It does not try to speculate about futures we cannot predict, nor is it alarmist. It is based on the best facts we can find about the here-and-now and the proven heritage of natural life forms and environmental conditions. It is optimistic about the future, based on humanity's past ability to progress, despite occasional set-backs. It supports 'blue-sky' thinking.

In the Green world wind farms are the angelic white saviour of the planet. In the Blu world, they are a failed attempt to resurrect a 13th-century technology with Green 'spin', but they do not spin enough to be at the core of a 21st-century mass-energy system. They are an unwelcome, unnecessary and costly add-on to essential gas-fired and nuclear-powered reliable electricity generation in wealthy countries, and a cruelly imposed barrier to the progress of developing countries which can only afford coal generation.

In the Green world, biofuels are somehow getting us back to a more natural way of using the planet's resources. In a Blu world, they are literally a waste of space and a threat to the optimisation of land use for agriculture, forestry and species-rich habitats. Large-scale wind and solar farms are also wasteful of space, about 500 times less intensive than gas and nuclear, never mind their sterilisation of potentially useful habitats and their danger to wildlife coming into contact with them.

Green activists waste their time searching for villains while Blu enthusiasts seek opportunities for improvement. The former seek alleged faults in Western democracies while the latter promote their patent strengths. The former are pessimists and the latter optimists. The former are so focused on their dictatorial narrative, they miss the harm their ideology is doing to the environment, Western economies and the geopolitical balance of power. The latter seek a holistic approach that strengthens and promotes liberal human freedoms, such that they inspire and enlist volunteers who will improve both their own lives and the condition of the environment around them. The former apply recycling as an ideological imperative whether it makes economic sense or not, while the latter only recycle where it is cost effective and sensible.

In the Green world, we need to completely overhaul our comfortable, safe and fairly happy modern lifestyles to be a slave to utopian concepts like 'saving the planet'. In the Blu world we take careful, small steps to make the world around us slightly better in a pragmatic, incremental way that will build momentum as we combine our efforts. In the Green world, we dramatise problems and seek radical options, which are guaranteed to fail in a big way. In the Blu world, we keep a cool head, analyse the problems using verifiable data, seek proven methods of addressing them, while keeping an open mind, knowing that solutions will inevitably have to be adapted as we go, because no one can predict for sure how they will work out.

As a true Blu rather than a Green I will not say we have 12 years to save ourselves and the planet, but the sooner we reject alarmism and adopt pragmatism, the better.

APPENDIX
REVIEW OF UN IPCC ASSESSMENT REPORT 6

BACKGROUND
This assessment refers to previous sections of the book but can be read as an essay on its own.

Due to the lack of empirical science in climate change research, an alternative approach is adopted where hypotheses are generated and, on the basis of certain observations, the chances of these being valid are assessed from "unequivocal" to "low confidence"; that is, various perceived levels of certainty/uncertainty are defined. Computer models of how the world's physical and atmospheric systems work are also produced. This approach has limited success – models generally have a poor record of replicating the real world – because various natural climate mechanisms, particularly associated with the oceans and the clouds, are poorly understood and they interact in a 'chaotic' way.

Even hypotheses which the IPCC expresses 'low confidence' in are relished by extremist environmental groups whose influence on democratic governments (the only ones which listen to them) leads to proposals for radical changes to our comfortable and affordable way of life, but which are inefficient at significantly reducing human emissions. If implemented worldwide, these energy policies would deny poorer countries the chance to share the lifestyles that most in the West currently enjoy. Even 'medium confidence' hypotheses are not good enough, given what is at stake. The UN needs to get its house in order.

GENERAL

Assessment Report 5 was my first introduction to the IPCC interpretation of climate change, which spurred me on to learn more, and has given me the knowledge to review Assessment Report 6.

Climate Change 2021: The Physical Science was published on 9 August 2021, the first of four sections of Assessment Report 6 (*IPCC AR6*), the others being *Impacts, Adaptation and Vulnerability*, *Mitigation of Climate Change*, and *Synthesis Report*, released in 2022. Like its predecessors, the first, the product of Working Group 1 (WG1), is far from reader friendly. The cynic would say this is because the IPCC does not intend it for the general reader, the political message in the Summary for Policymakers, the first 41 pages out of 3,949 total, being the prime communication device of this political organisation. Nevertheless, in trying to separate the politics from the science, and identify the political direction of travel, examination of the whole document would be necessary. Unfortunately the body of the report is marked "Subject to final edit. Do not cite, quote or distribute". This effectively censors public criticism, and why say that outsiders should not distribute the report when it has made it downloadable? This review is therefore confined to the Summary for Policymakers, which is generally referred to in this book as "the Report".

I can understand the IPCC editors' reluctance to go nap and expose themselves to more detailed criticism. Pumped up with confidence in the belief that their authority is unimpeachable, and intoxicated by groupthink, pride can go before a fall. *IPCC AR4* (2007) had some very embarrassing mistakes. Donna Laframboise reviewed it in her book *The Delinquent Teenager* (the title was a reference to the IPCC as an unruly teenager) and concluded that a third of the content had not been peer reviewed, such that it would have obtained Grade F if submitted to a university. The Netherlands was said to be 55% below sea level when the correct figure was 26%. Antarctic ice was underestimated by 50%. Most famous of all was the prediction that Himalayan glaciers would melt by 2035. This was so ridiculous that the IPCC issued a statement saying that perhaps it meant 2350. And, of course, there could be an "ice-free Arctic Ocean for a short period in summer perhaps as early as 2015". Realising this was not going to happen, the UN stated in 2014 that "By 2020 one would expect the summer sea ice to disappear." Its predictions in this respect have now been tempered, the Report pushing the date out to at least 2050 in all its scenarios in Figure SPM.8(b). It could be some time until the whole truth about *IPCC AR6* comes out. It took until a whistleblower leaked the 'Climategate' emails in 2009 to expose the jiggery-pockery which led to the hockey-stick graph of historic temperatures in the 2001 *IPCC AR3*. This graph returns in the Report as Figure SPM.1(a).

I am constantly conscious that the 'climate scientists' who compile and edit a report like this (3 review editors, 15 coordinating editors, 71 authors and 34 contributing editors – 123 in total) have clear objectives set by the IPCC. They are not the applied scientists who do useful work like invent vaccines for COVID-19, but are academics who observe things and construct models of how they think the real world behaves. Empirical, solid science is sadly lacking in climate change research. Each assessment report contains a series of contentions and opinions as to the likelihood of them being correct, from 'low confidence' to 'unequivical'. As we have seen in previous chapters, the mood of the academics carefully chosen by the IPCC has developed from the first report which pointed out that, since we started to emerge from the Little Ice Age in the 19th century when human emissions were comparatively small, there must have been strong natural climate warming factors at work, until now when natural climate change is effectively discounted.

The IPCC is a political organisation (the Intergovernmental part of its title makes this clear) where certain prominent politicians, mostly in the USA and the UK, have not been content with making sure that potholes in the roads are filled or the local hospital gets the latest cancer-diagnostic equipment, but who wish to 'save the planet'. There are major climate mechanisms we still cannot explain, like how an El Niño absorbs the heat from the Sun and releases it in great bursts that can vary the temperature of the whole world by more than 1°C over as little as a five-year period, or whether changing cloud patterns and quantities provide a positive or negative feedback to climate change. In 1992, Roger Revelle, who first alerted Al Gore to the possibility of anthropogenic global warming at a class at Harvard, declared that scientists needed perhaps another 20 years of research before they could come to conclusions, but by the third assessment report in 2001 the political pressure on its editors to perform led to humanity being firmly placed in the dock, a direction of travel that has continued to the current, sixth report. Advisors who were not on board with this mission have been dropped along the way.

When I previously reviewed the CVs of the 19 authors of the *IPCC SR1.5* report (2018) there were few of them, if any, I would describe as eminent scientists. The sum total of practical experience of a so-called expert in oceanography was, "Fieldwork: nine weeks at sea across fishing boats, research vessels and a tall ship". Sounds more like a publicly subsidised jolly to me. I note that Sarah Connors is an author in both reports, who only gained her PhD in atmospheric chemistry in 2015, and called herself a project manager in science communication despite appearing to have no significant scientific research experience. And yet activists claim that the world's top scientists work for the IPCC. Heaven help us if these are the best on offer. I concluded that such people were chosen by the UN for their ability to deliver the desired message and their opportunity to achieve credibility by having an IPCC document on their CVs. Lead author Valérie Masson-Delmotte is keen to sort the gender imbalance by including more women, perhaps explaining why Sarah Connors was enlisted only a few years after she obtained her PhD. Competency and experience seem to be of secondary concern.

The 123 contributors are from 36 countries, with many claiming to represent more than one country, perhaps to get the figures up so that the global contribution appears equal to the global challenge. Shubha Sathyendranath claims to be from the UK, Canada and India (she works at the Plymouth Marine Laboratory in the UK on the ocean carbon cycle), and Irina Gorodetskaya is from Portugal, Russia and Belgium (she works in Portugal at the University of Aeiro on atmospheric processes and modelling).

Paleoclimatology expert Masson-Delmotte was the lead author of *IPCC SR1.5*, and she is one of the three review editors of *IPCC AR6*, the others being Greg Flato from the University of Victoria in Canada, and Noureddine Yassa from Morocco. Before publication, Masson-Delmotte and Panmao Zhai were the co-chairs of WG1, but he is not mentioned in the final report. Since 2010, Zhai has been Secretary General of the Chinese Meteorological Society, a Chinese Communist Party appointment, but I could not find out from the web if there was anything sinister involved in his omission. The view from the West regards the lack of serious participation in emissions reduction in China as a desperate measure by Xi Jinping to keep his oppressed people sweet by maintaining low energy costs but, as I pointed out in Chapter 12, some Chinese scientists, free from the political influence of the West-centred IPCC, have demonstrated the large natural variations in climatic temperatures over the last 10,000 years, contrary to the hockey-stick in Figure SPM.1(a). Perhaps the Chinese are merely following their science, while paying lip service to the West's climate change agenda to continue to sell us renewables.

Yassa is also a government appointment (Director of the Centre for Development of Renewable Energy) in a totalitarian state which comes 96th in The Economist Intelligence Unit's ranking of democratic rights and civil liberties. According to a report in *Energy World* on 29 January 2019, 98% of the country's power output of 19,000 MW is gas generated, with Yassa trying to launch a single solar farm with a capacity of 150 MW in this impoverished country. In its attempt to embrace a one-world approach to the 'climate crisis', the UN is not fussy who it deals with, neither China nor Morocco exactly being the model advocates of 'net-zero', nor will they care if the expensive pursuit of the latter unachievable objective harms Western democracies.

A key conclusion of the Report is that:

- Current warming is unprecedented in many thousands of years.
- Therefore practically all the warming since 1850 must be due to human emissions.
- Therefore the sensitivity of natural climate mechanisms to human emissions must be large.
- Therefore the future warming rate will be scary; and
- Once fed into climate models, we are definitely wading into the brown stuff.

It is paleoclimateology (the study of historic climate through proxy data like ice cores, sediments, tree rings, etc.) that is used to justify the unprecedented global warming hockey-stick graph, the starting point for this piece of logic.

At the time of the COVID-19 crisis, key government scientists were exposed to the scrutiny of the media at press conferences and had to perform. Why are Masson-Delmotte and her two colleagues not similarly under the spotlight, apart from the fact that they might crumble under the searching questions of a good interviewer? In the absence of being able to personally interview them, I have included quotes in the analysis below from Masson-Delmotte and Flato (Yassa seems to communicate only in Arabic and French on the internet and my command of both is limited), which reinforce my contention that:

- The science is far from settled.
- Computer modelling is still unfit for purpose.
- We do not currently have the means to reduce emissions by a significant amount, certainly not to net-zero.

The tactic of the UN is to keep such scientists protected in background anonymity so that a politician like António Guterres can take the spotlight, supply the added rhetoric, and declare "a code red for humanity".

I am interested to find out what the Report says about the following:

1. What is its analysis of the current state of warming?
2. Is it still contending that current temperatures are unprecedented, ignoring evidence that they were higher at various points during the Holocene?
3. How much of the post-1950 warming is now attributed to human factors, and has any progress been made in better defining the transient climate response to increased CO_2, or is it still no better than Jule Charney's assessment in 1979 of between 1.5°C and 4.5°C? What is the evidence for a long residency period of CO_2 in the atmosphere?
4. Is there any attempt to explain why the temperature stopped rising in the periods 1945–75 and 1998–2014, which suggests a poor correlation between rising emissions and global warming?
5. Is any regard paid to natural climate mechanisms and the research which has been carried out on these since the last assessment report in 2013?
6. Has the IPCC finally fallen into the trap of citing current extreme weather events as proof that catastrophic global warming is taking place?
7. Is the IPCC still admitting that climate models are effectively not fit for purpose in terms of predicting climate change?
8. What is the quality of evidence on sea level rise and ocean acidification?
9. How well is its scare strategy developing, with particular reference to the 1.5°C target?
10. Is it any more convincing that there is a practical, never mind affordable, way of radically reducing emissions?

I often make comments about what the Report does not say, which is just as relevant as what it does.

1. WHAT IS THE CURRENT STATE OF GLOBAL WARMING?

The planet has warmed by an average of 1.09°C (±10%) from the temperature in the period 1850–1900 (para. A.1.2). After the pause in warming from 2003 to 2012 (I would argue from 1998 to 2014) there was warming at the end of the 2010s, and "methodological advances" have also contributed to the stated warming since the assessment in *IPCC AR5* in 2013. Such changes in the assessment method, rather than observations, are often described as 'recalibration' of data, and it is interesting how each time this is done by the main government agencies fronting climate change research (the UK Hadley Centre and the US NASA and NOAA) the warming increases. It never decreases. Surface temperature data collection has deficiencies such as a lack of observatories in various parts of the world and urban heat island distortion in non-rural observatory locations, but my preferred satellite-based observations (since 1980) say something similar, so it is safe to assume that the planet is about 1.1°C warmer than the middle of the 19th century, with land areas slightly warmer on average than ocean areas, and the northern hemisphere slightly warmer than south of the equator because there is more land there.

The Report (para. A.1.5) attributes the decrease in Arctic ice between 1979 and 1988 and between 2010 and 2019 (what happened between 1988 and 2010?) to "human influence", although I argue in Chapter 12 that the 60-year cycle of the Atlantic Multidecadel Oscillation (AMO) plays a key role, since satellite data from 1964 to 1979 showed an increase in ice extent when the North Atlantic was in a cooling phase (but when human emissions were rising in a post-WW2 industrial boom). The IPCC never mentions this. It says: "In 2011–2020, annual average Arctic sea ice reached its lowest level since at least 1850 (*high confidence*). Late summer Arctic sea ice area was smaller than at any time in at least the past 1,000 years (*medium confidence*)." As I showed in Chapter 12, the low point in Arctic ice in the 1950s, when the nuclear submarine USS Skate surfaced at the North Pole, was similar to that in 2012, when the extent of ice was at its lowest in the current AMO cycle, but still at 3,600,000 km² in September that year. It was at 5,400,000 km² in September 2014, the year that Al Gore predicted in 2009 it would be all gone. It varies a lot each year with the trend since 2007 unclear, but if anything upwards.

The Report predicts warmer times for the Arctic: "The Arctic is projected to experience the highest increase in the temperature of the coldest days, at about three times the rate of global warming." We have relatively poor historic records of temperatures above latitude 65°N because there are so few human settlements there and it is impossible to place weather stations on floating ice that constantly moves. In Chapter 12 I did the next best thing and plotted temperatures at Reykjavik from their beginning in 1940, when the average annual temperature was just over 5°C (boy, do the Icelanders need global warming!). It dropped to less than 4°C by 1979 (AMO influence?) when the IPCC started to quote figures for Arctic ice. By 2010 it was back up to over 5°C, only about 0.25°C warmer than 1940. This would suggest that the current cooling phase of the AMO towards 2040 will temper any rise in general global temperatures in this area, with consequent slowing of any Arctic ice decline. But to predict this would be to fall into the same trap that the IPCC does, pretty well all climate predictions having been wrong in the past because, I contend, the 'chaotic' nature of complex climate mechanism interactions is beyond our ken.

Like *IPCC AR5*, the Report concludes that "there has been no significant trend in Antarctic sea ice area from 1979 to 2020", and "there is *low confidence* in the projected decrease of Antarctic sea ice", so the alarmists like David Attenborough who say that cute penguins are at threat from increasingly fractured ice (BBC's 2019 *Seven Continents, One Planet*) have not read the science. He also gets it wrong with polar bears, which increased in number almost sixfold between 1979 and 2020, a true conservation success story, not a disaster caused by greedy, gas-guzzling humanity.

The IPCC has better defined what it means by pre-

Industrial Age temperatures by comparing global warming to the average in the period 1850 to 1900. However, we should be cautious about obtaining figures for a period of time without reliable observations in most parts of the world. It becomes even more incredible when the Report tracks precipitation and soil moisture change from a baseline in the middle of the 19th century (Figure SPM.5). No one has been measuring these factors over the whole of the world for 170 years from which to model the effects of more warming. The figures involved can only be conjecture.

2. ARE MODERN TEMPERATURES UNPRECEDENTED?

This is the key aspect on which the scary message of the Report depends, illustrated by Figure SPM.1(a), a hockey-stick graph with a scary upwards tick. It is the million-dollar question, or rather it is the hundred-trillion-dollar question if it is believed that current and impending temperatures are unprecedented, and governments are persuaded to spend a fortune on (vainly) attempting to achieve net-zero emissions. "The scale of recent changes across the climate system as a whole and the present state of many aspects of the climate system are unprecedented over many centuries to many thousands of years. The Last Interglacial, around 125,000 years ago, is the next most recent candidate for a period of higher temperature. The last time global surface temperature was sustained at or above 2.5°C higher than 1850–1900 was over 3 million years ago (*medium confidence*)." Just in case you think that "many centuries" or "125,000 years ago" is not scary enough, it raises it to "3 million years ago" for good measure. Plenty of quotes there for the alarmists.

In Figure SPM.1(a) the Medieval Warm Period is less than a quarter of a degree warmer than the 1850 datum, and the Little Ice Age less than a quarter of a degree colder. I find this incredible. Surely half a degree cannot be the difference between freezing rivers in one era and vineculture in another in southern England? Surely the Vikings could not have lived in Greenland from 985 to 1406 if it had been three quarters of a degree colder than present, and they could not have buried their dead in ground which is currently in permafrost, or have sailed round the island, a feat which is still not possible today?

Let me summarise my alternative hypothesis. The world has warmed by an average of just over 1°C since the beginning of the 19th century, the end of the Little Ice Age, a period when the River Thames often froze over in winter. This means we are part way back to the warmth experienced in the Roman and medieval periods when commercial vineyards grew in England for many centuries, a feat only possible in this warm phase for the last 40 years. It is possible, even likely, that most of the warming in the last 200 years has been natural. Increased CO_2 must be making a contribution but we do not know enough about the mechanics of climate change to accurately determine its transient climate response. However, the measured warming over the last half century suggests the bottom end of the Charney Range. This is nothing to worry about in the short term, but in case the temperature starts to creep over about 2°C above that of 1850 we should explore low carbon methods of generating power for possible mass implementation later this century. At present, some version of nuclear power seems most likely. In the meantime, more atmospheric CO_2 is greening the planet. So far, so good.

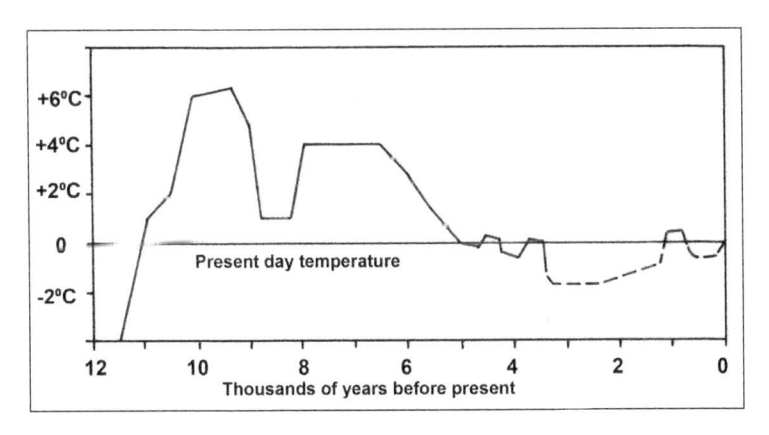

Figure 41 Proxy temperatures at Svalbard.

In Chapter 12 I presented historical evidence of civilisations prospering in warm and wet periods in Bronze Age, Roman and medieval times, and coming to grief in periods that were colder and drier. I also cited several proxy data studies which indicated that temperatures peaked in the Holocene (the period since the last main Ice Age) around 9,000 years ago, when the Arctic was at least 4°C warmer than present (see Figure 41). *Mangerud and Svendsen* 2017 charted the remains of marine moluscs in sediments at Svalbard, previously called Spitzbergen, a Norwegian archipelago centred on 78°N, to determine how warm or cold the Arctic was compared to today. Blue mussels made a brief appearance during the Medieval Warm Period, if not the Roman one, so the lack of proxy evidence on the latter explains the dotted line from about 3,000 to 2,000 years before present. Arctic ice is currently at a high point within the Holocene (see also my Figure 36 in Chapter 12). The IPCC started to chart it in 1979 (despite having satellite data since 1964), and other data suggests a peak at this time on a 60-year cycle. Polar bears have survived in much warmer conditions than currently exist, and they have increased in number from about 5,000 in the 1970s to almost 30,000 now in a time of reducing ice. Yet they are supposed to be endangered.

In Chapter 12 I also charted the invention of the hockey-stick temperature graph, citing the leaked emails of those involved to show that cherry-picked tree ring data (with flat line temperatures, even up to the late 20th century) was spliced together with measured temperatures from 1960 to suggest that modern day warming was unprecedented in 1,500 years, an objective that the emails made clear had been set by political masters.

IPCC AR5 was content to state that most of the temperature increase since 1950 was due to human factors, but the Report now says that practically all of it was, even from 1850. If the warming is all down to emissions, this suggests a high transient climate response to CO_2, and a speedy advance towards 2°C above pre-industrial temperatures. Start panicking now.

3. HOW MUCH WARMING IS DUE TO HUMAN EMISSIONS, HAS THE TRANSIENT CLIMATE RESPONSE BEEN BETTER DEFINED, AND WHAT IS THE RESIDENCY PERIOD OF ATMOSPHERIC CO_2?

In almost every paragraph of the Summary for Policymakers, human factors are stated to have caused every aspect of climate change that is covered, as if constant repetition of this assertion will make it more true. It must all be our fault, so we must take immediate action to curb our greed for fossil fuels.

If the climate has warmed by about 1.09°C, the Report attributes 1.07°C of this to human causes (para. A.1.3). The last two assessment reports had merely said that most of the warming since 1950 had been due to human factors, so this is a big change. The implication is that the climate would be very stable without human interference, and that it is very sensitive to our emissions. The graph in Figure SPM.1(b) is compelling, but it should be noted that the "natural only" flat line "simulation" only takes into account the tiny variations caused by solar radiation changes and volcanic activity. There are many more natural climate mechanisms being studied by scientists the UN refuses to acknowledge although, as I point out in Chapter 9, those scientists are candid enough to admit that they are only scratching at the surface because climate science is still at a very basic level of understanding.

Since 1979, despite decades of research, the wide Charney Range (1.5°C to 4.5°C) has been our best estimate of what might happen if we doubled atmospheric CO_2 concentrations. The Report assumes a higher climate sensitivity from *IPCC AR5* by amplifying cloud positive feedback by about 20% and, with no best estimate in *IPCC AR5*, returns to a median Charney: "best estimate is 3°C with a *likely* range of 2.5°C to 4°C (*high confidence*)" (para A.4.3). I get frustrated by IPCC-speak. It "likely" has "high confidence". The definition of cloud feedback is a key factor. If humans start to warm the planet there should be more water evaporating and water vapour is the main greenhouse gas. So do more clouds shield the sun (a negative, cooling feedback) or increase greenhouse warming (a positive, warming feedback)? With their inveterate, pessimistic mindset, IPCC contributors tend towards the latter, adding on 20% from the last assessment report for good measure, but in truth no one knows for sure how negative and positive feedbacks balance.

During the Paris climate talks, Valérie Masson-Delmotte was asked about the likely value of climate sensitivity: "I cannot say with the precision of half a degree. I think, so far, from what I have seen, there has not been any comprehensive study that would allow [us] to produce an informed number. I have a feeling – it's a feeling – that probably values between around 2.5°[C] or more may be more realistic" (YouTube *Valérie Masson-Delmotte on the likely value of climate sensitivity*). So one of the top three experts in the Report had to rely on her feelings to guess the transient climate response, there being no "comprehensive study" (why not after a quarter of a century?), and her guess was below the median of Charney.

Greg Flato, who majors in earth system models, was a contributor to a paper in *Science Advances* on 24th June 2020 which discussed two approaches to determining the outcomes of doubling CO_2 emissions, the equilibrium climate sensitivity (ECS) approach, which has a range of 1.8°C to 5.6°C, and a transient climate response (TCR) approach, which points to around 1.7°C but with a possible range of 1.3°C to 3.0°C. If this is what he said the previous year, and he is the Earth systems model expert reviewer on the IPCC, why has he allowed the Report to say something very different? The range 1.3°C to 1.7°C is similar to the conclusions of people like John Christy, Richard McNider, Nicholas Lewis and Judith Curry, scientists who were dropped by the IPCC some time ago, and who used the Energy Balance Model (see Chapter 8) which majors on analysis of past observations, not speculation of possible future trends. The bottom of the Charney Range does not present a scary scenario, just an issue which needs to be further reviewed going forward, with research into lower carbon energy sources in case we need them.

Incidentally, Flato was one of eight authors of this small paper in *Science Advances*. It is as if academics need to group together for safety, frightened that any one person is blamed if something is proved wrong. Does each person not have sufficient courage of his convictions to put his head above the parapet? Is this a convenient way to mount up the academic paper count to enhance one's reputation among other academics and attract further funding? Is it good value for public money for eight people to carry out a task instead of one or two? Are *Science Advances* made by merely reviewing other people's studies, and coming to conclusions that we already know, that "cloud feedbacks and cloud-aerosol interactions are the most likely contributors" to the wide differences in figures for the transient climate response? At least people like Henrik Svensmark and Nir Shaviv (Chapter 9) are carrying out experiments to better understand how clouds affect the climate, but they are ignored by the IPCC because they are liable to demonstrate the part that natural forces play in current climate change, a concept that the IPCC must avoid at all costs.

The UN is a political bureaucracy with the admirable but naïve aim of saving the planet and its inhabitants from perceived problems. Its analyses have certain features. They tend towards pessimism to emphasise the problems, seeking to obtain enthusiasm and support to address them. While various scenarios are considered, opinions coalesce towards a status which might be referred to as a 'scientific consensus' even if it is more speculative than empirically substantial. Once a consensus becomes entrenched it is difficult to revise as new evidence appears, and it is impossible to challenge because the bureaucracy behind the consensus is anonymous and out of reach. To determine the current reality, it is better to critically assess prominent individual scholars (the sceptics), even if they are ostracised by the establishment, because they have the freedom and bravery to think outside the box.

Consensus is a curious thing. I once heard a journalist say that he was part of a small group which wrote the newspaper's editorial each day. It had to determine what the title's official stance on a particular subject was. Sometimes he concluded that the common consensus reached was not the view held by any of the contributors, but an amalgamated compromise. He felt uneasy about this because it did not seem real or honest.

In trying to predict future population trends or climate states, the complexity and multifariousness of cultural, meteorological or ecological systems are grossly simplified to draw politically clear conclusions which are far from reality. This explains why all COP conferences have been complete failures.

One of the key propositions of my book is that the academic world values group working and collegiate consensus, which can lead to groupthink and a politeness that fosters political correctness. The commercial world I spent much of my working life in often rewards the individual who thinks outside the box in a very radical and impolite way, and constantly challenges the status quo to arrive at a successful unique selling point from among the multitude of daft ideas, serendipitously thrown up, that are rejected by commercial reality. In their more protected and cloistered world, academics are less exposed to the harsh realities of the evolutionary process that drives human progress, with its slings and arrows of outrageous fortune. I argue that the scientific process is impaired by the niceness and compliance of academics which explains why a group document like the Report, designed by committee, is so hopelessly flawed (in my analysis).

The UN IPCC might as well put up a sign saying: 'Mavericks need not apply since consensus is the political imperative'. It is easy to find 123 academics who will go along with this, whether they are conscious of doing so or not. My thesis is not one of conspiracy, but of nice but fallible human beings reacting to an ostensibly noble cause within the prevailing mindset and the strictures of the current political landscape, particularly in Western democracies which are rich enough to take time out to have a crisis of confidence of many of the things that have so far driven human progress: fossil fuels, nuclear energy, capitalism, big business, plastics, intensive agriculture, GM technology, etc.

In an interview, Valérie Masson-Delmotte admitted that it was impossible to keep up with the 40,000 academic papers churned out each year with the words 'climate change' in them. Critical evidence could be missing from the IPCC's deliberations; for example, the ones I have cited which present evidence that current temperatures are not unusual or that show that natural climate change mechanisms are present. Since papers generally make reference to, and rely on, other people's work, they are usually reviewed by sympathetic peers (we have seen that peer review is a very poor quality control mechanism), and 97% of papers refer to climate modelling in some way (and models are usually wrong), there is ample opportunity for errors and misconceptions to spread like a contagion (I call it CO_2VID). The consequence of such discrepancies in academia is merely increased polite debate and wasted public funds, while in the commercial world they would lead to bankruptcy and ruin. Why are there so many papers produced despite left-of-centre politicians like those mentioned in Figure 42 below who initiated the climate alarm and have been saying for decades that "the science is settled"? Who says that the science is not following the politics?

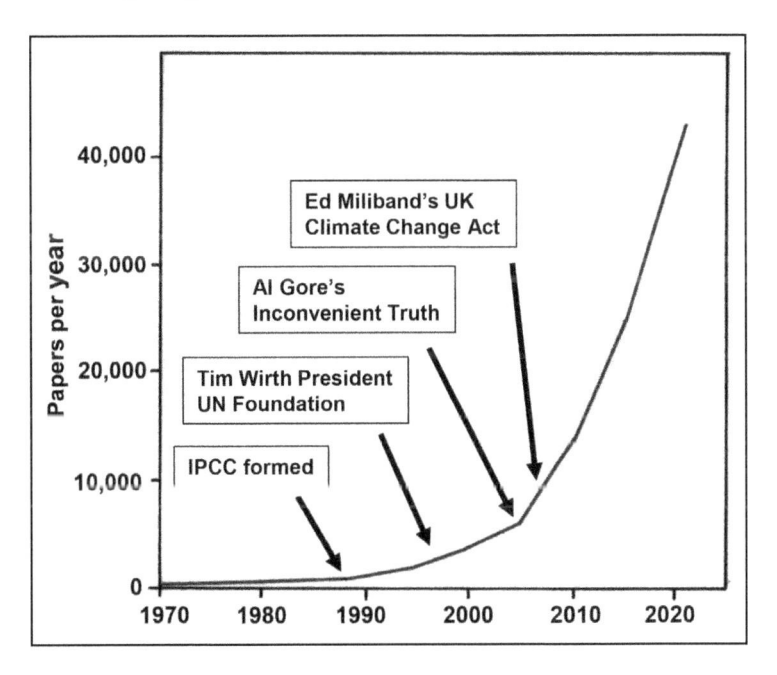

Figure 42 Number of academic papers per year with the words 'climate change' in them – a hockey-stick graph.

THE ANTIDOTE TO ECO-ANXIETY | 307

If we are calculating how much CO_2 there is in the atmosphere we need to know how long historic emissions will stay there (the residency time) and accumulate, rather than be absorbed into the oceans and used up in the photosynthesis process. Before the 1990s, atmospheric CO_2 was believed to have a half-life (the time it reduced by half) of about ten years, but the Bern Model (yet another model of how the world might work) now assumes 29 years, with *IPCC AR5* declaring that large amounts of CO_2 will remain in the atmosphere for many centuries. The Report does not mention what residency time is assumed.

4. WHY HAS THERE BEEN A POOR CORRELATION BETWEEN RISING EMISSIONS AND GLOBAL WARMING?

IPCC AR5 devoted three pages to try to explain what it called the "hiatus", the pause in warming from about 1998 to 2013. I have taken information from the UK Hadley Centre Hadcrut5 model to compile Figure 43. It could be said that I have been somewhat mischievous highlighting the temperatures at the 2001, 2007, 2013 and 2021 dates when IPCC assessment reports were published to show that they all lie close to the hiatus line on the graph, with the small rise from 2001 to 2021 consistent with a 0.13°C per decade long-term trend. However, this is no more mischievous than Figure SPM.1(b) in the Report which takes the peak of the period 2015–2020 as proof that global warming is currently as virulent as it could be. Such a short period as this can only be interpreted as weather.

To the climate pessimist (humans are causing a climate crisis) the 2015–20 peak in temperatures is proof that they are right, although they still cannot explain the 1998–2013 pause. To the climate optimist (humans play a minor part in predominantly naturally occurring climate change) the hiatus and the peak are both determined by natural factors, the latter triggered by the very large 2015/16 El Niño event. We saw in Chapter 5 what devastation an El Niño can have, both locally and globally, the difference between a La Niña low and an El Niño high being over 1°C average over the whole planet in as little as a five-year period. Locally, the 1939 El Niño caused a

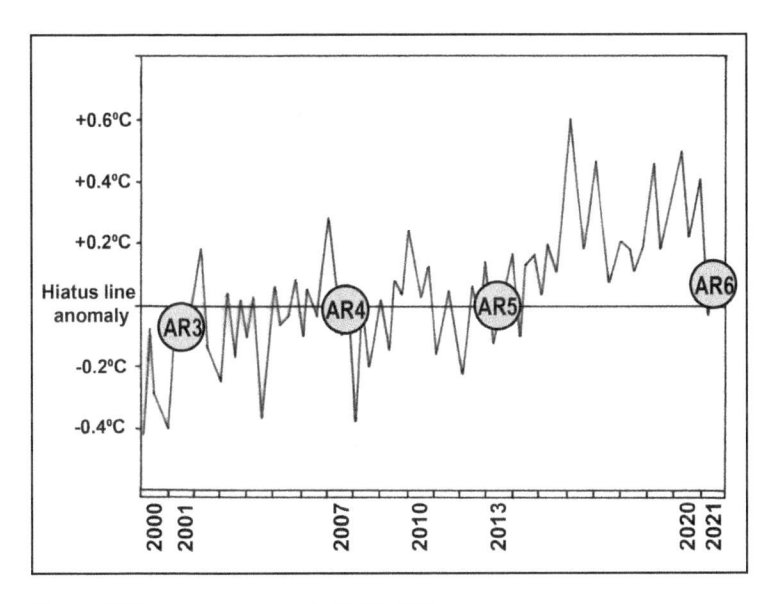

Figure 43 Temperature anomaly 2000–2021.

rise in temperature of 10°C in parts of the Humboldt current, with devastating consequences for wildlife, and the 2015/16 El Niño caused a huge area of coral bleaching in the Great Barrier Reef, from which it has subsequently fully recovered.

The Report says nothing about the fall in average world temperatures from about 1945 to 1975, when post-WW2 economic recovery emissions were rising, and many scientists thought we might be returning to the Little Ice Age temperatures which had been the default state for 400 years up to about 1820. But then, the climate pessimists of today do not recognise that the LIA existed. Instead the Report turns the evidence upside down by stating in paragraph D.1.1: "This Report reaffirms with *high confidence* the AR5 finding that there is a near-linear relationship between cumulative anthropogenic CO_2 emissions and the global warming they cause." Why then is the warming trend from 1850 not an upward-curving line (Figure SPM.10 is an almost straight line), reflecting the cumulative emissions which are alleged to reside in the atmosphere for a long time? The discrepancy between the logarithmic increase in emissions and the straight-line increase in temperatures can be explained by each new quantity of CO_2 emissions having a smaller greenhouse radiative effect than the one before, as the radiative effect tends towards saturation point. Certain radiative wavelengths that CO_2 (and H_2O) operate at are already saturated, such that more atmospheric CO_2 will make no difference to temperatures at these ranges. But stating this might moderate the alarm that the IPCC is committed to uphold.

5. WHAT ABOUT THE MECHANISMS OF NATURAL WARMING?

While we can draw a straight line from 1850 to today and conclude that the world has warmed, it has done so in a stepped fashion which does not correlate with human emissions production, with periods of cooling (c1880–1915 and c1945–75) as shown in Figure 7 in Chapter 3. The Report recognises this and expects it to continue, but it briefly dismisses it as "internal variability" (para. C.1.1 to C.1.3), confirming that it cannot adequately explain it.

We do not understand the mechanics of ocean features like El Niños, which have effects on weather patterns beyond the Pacific area it originates from, or the Atlantic Multidecadel Oscillation which seems to have a cyclical effect on Arctic ice extent. The poor understanding of clouds and their effects on ocean currents, solar radiation shielding and scattering of greenhouse gas radiation also greatly adds to climate change uncertainty.

The IPCC's original remit was to explore how humans are affecting climate so the vast majority of research funding has gone on issues associated with anthropogenic factors. The Report confines itself to the effects that volcanos and direct solar radiation have on climate, which are small and temporary. There are other theories it avoids, not just because they are not yet fully understood (that has not stopped it including half-baked "low confidence" data before), but perhaps because it cannot contemplate dilution of the blame placed on humanity.

People like Henrik Svensmark are exploring other factors. While the Sun's direct radiation does not vary enough to explain the ups and downs of temperature over the last 170 years, he is interrogating how changing solar gravitational forces allow different amounts of cloud-seeding cosmic rays to enter the atmosphere. A one per cent change in cloudiness over years or decades would have a bigger effect on temperatures than the assumed radiative power of human-produced greenhouse gases.

6. IS EXTREME WEATHER BEING CONFUSED FOR CLIMATE CHANGE?

There does not seem to be a day go by but news reports highlight an extreme weather event, especially in the run-up to a COP event. I have not found much evidence to show that extreme weather events are statistically increasing over the several decades that we must review to distinguish climate from weather, but I can be sure that news agencies relish the drama of people struggling against storms and wildfires. While we are used to seeing activist groups get hot under the collar about the changing climate, the general public is more concerned about whether they will get wet tomorrow on a day out at the seaside. They would prefer to take the suntan lotion than raincoats. Climate pessimists fervently and honestly believe that they must get a more immediate message across to the public so they look for evidence that climate change is having harmful effects on today's weather, even though they know that no good scientist would try to prove, and therefore confuse, climate with weather.

The Report capitalises on the public appetite for bad news by majoring on three aspects: hot extremes, heavy precipitation and agricultural and ecological drought. The world is getting warmer, so it is easy to find record high temperatures since 1850, and common sense tells us that summer heat waves should be warmer. But that also means that winters are getting milder, a thoroughly good thing because we know that very cold winters cause almost 20 times more deaths than very warm summers (*Lancet* 2015 study of 74 million deaths). The growing season is increasing by two days per decade. What farmer, or anyone else for that matter, would not prefer longer summers and shorter winters? During my lifetime, summers have become two weeks longer and winters two weeks shorter. That is slow change but it is wonderful. Perhaps it is because we evolved as a species in tropical Africa that we predominantly go to warmer areas on holiday, and there is a net migration of Americans to California, Texas and Florida, despite the heatwaves and tropical storms that are occasionally encountered. We have learned to adapt to warm climates and weather extremes.

A warmer atmosphere should also mean that cyclones hold more water and we should get heavier downpours, but this just means that many areas will have more water to capture, if we invest in the infrastructure to obtain this and to prevent flooding, an activity we have proved ourselves very good at over the years. In Chapter 12 I showed that past civilisations prospered in warmer, wetter periods because agricultural output improved.

It is difficult to predict the weather more than a few days in advance, so predicting weather patterns several years or decades in advance can be no more than speculation, educated guesswork at best. The authors of the Report divide the world into 45 regions (we are all in equal hexagons – heavens, please tell me this is not the UN's idea of a new political world order inspired by John Lennon's "imagine there's no countries"). It tries to predict how climate change will pan out, and it gives each regional assessment a figure for the level of confidence that the prediction will transpire.

It is pretty confident that hot extremes will generally increase, the only areas where the jury is still out being Central and Eastern USA (was that a political decision to omit them?). However, the only areas it has any confidence that heavy precipitation events will increase are Central USA and northern Europe. I am sure that the farms of Dakota, Nebraska, Kansas, Iowa, Minnesota, Missouri and Oklahoma might like the choice of whether to utilise the extra rain or let it run away into the Mississippi, and speaking as a resident of northern Europe, I know how to cope with heavy rain.

When it comes to agricultural and ecological drought, the only two areas it has confidence in predicting this problem are

Western USA (presumably California, Oregon and Washington states) and the Mediterranean. There are lots of people in these areas who are enthusiastic supporters of climate change alarmism, or am I being overly cynical? They are certainly comparatively rich areas, and rich people have a habit of sorting problems.

In Chapter 1 I presented evidence that the area of the world in drought did not increase between 1982 and 2012, the area of the world experiencing wildfires has reduced in the satellite era (we are better at protecting our farmland and bushland around inhabited areas, except where green interests stop the maintenance of fire breaks because trees are seen as precious) and the planet is greening partly because of more plant-enhancing CO_2 in the atmosphere. The Report has some concern about possible drought in the Sahel, but the Leaf Index (the amount of vegetation) there has increased by more than 14% since 1982. The higher the concentration of CO_2 in the atmosphere, the less a plant has to open its pores to ingest it, and the less it releases moisture. Plants can withstand more drought because of more CO_2, but the IPCC is immune to good news. It describes this life-enhancing gas as "air pollution". The IPCC gradually assumes language like this from extreme climate alarmists, perhaps because it invites them to contribute to its political cause.

In a challenge to the contention that the current warming periods (c1975–1998 and 2015–2021) are particular times of increased extreme weather events, John Christy (ex-NASA and now University of Alabama in Huntsville) charted how often high temperature records were set (shown in Figure 44), with the c1920–45 warming phase being the highest. This does not necessarily mean that some of these temperatures have not been exceeded since, but that this was the time which saw the biggest change in extreme temperatures. Tony Heller found a similar result from his study of days with US temperatures over 100°F, the 1930s still being the record time for these, as witnessed by the Dust Bowl in the American and Canadian prairies between 1934 and 1940, where the drought, allied to

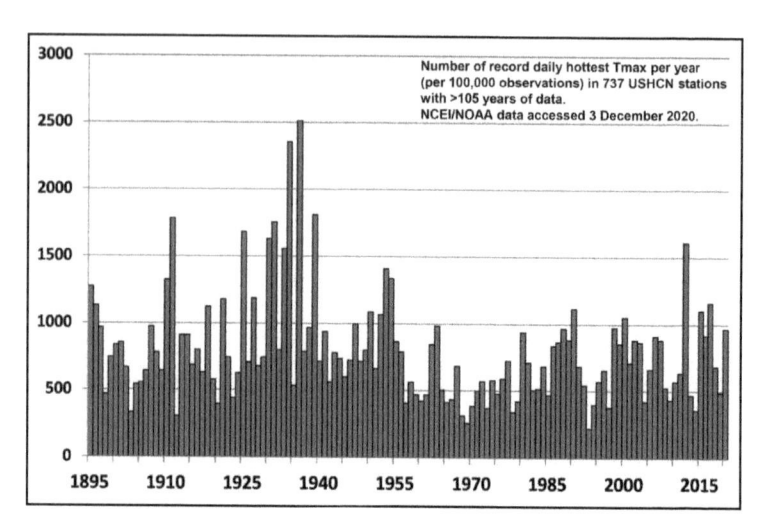

Figure 44 Frequency of high temperature records.

ill-advised ploughing methods which caused soil erosion, led to 3.5 million people being displaced. Today's extreme weather events are mild in comparison.

If polar regions are experiencing more warming than equatorial ones (although I have questioned this above with my analysis of Reykjavik historic temperatures), this means that the temperature gradient between the two is reducing, which should theoretically lead to less violent storms, but our poor understanding of oceans and clouds make it impossible to draw conclusions. Para. C.3.4: "The Atlantic Meridional Overturning Circulation is *very likely* to weaken over the 21st century". The Gulf Stream is part of this circulatory system and changes in it would affect the climate of Western Europe and the Eastern Seaboard of America, although who can say whether this would be for the better or the worse? When I checked the UK Met Office website, it noted that measurements of the AMOC had only started in 2004, stating, "It is too early to say for sure whether there are any long-term trends". But it seems that the IPCC cannot resist another opportunity for alarm as there will "*very likely*" be "drying in Europe", so it turns great uncertainty into "very likely".

7. ARE CLIMATE MODELS STILL UNFIT FOR PURPOSE?

Page 824 of *IPCC AR5* stated: "Climate models are extremely sophisticated computer programs that encapsulate our understanding of the climate system and simulate, with as much fidelity as currently feasible, the complex interactions between the atmosphere, ocean, land surface, snow and ice, the global ecosystem and a variety of chemical and biological processes... Climate models of today are, in principle, better than their predecessors. However, every bit of added complexity, while intended to improve some aspect of simulated climate, also introduces new sources of possible error... Furthermore, despite the progress that has been made, scientific uncertainty regarding the details of many processes remains." In other words, more complexity leads to more error, and there are key aspects of the climate we do not understand.

As I showed in Chapter 3, by citing the graph on page 981 of *IPCC AR5*, almost all official climate models have exaggerated the observed warming since 1990, most by a lot (see also Figure 45). This, and the fact that they cannot even "hindcast", that is, simulate past climate change, demonstrates that they are unfit for purpose. Eco-pessimists will contend that the warming between 2015 and 2020 takes the observed warming line up into the modelled zone in Figure 14, but this is too short a timescale to be relevant on its own. The Report claims that its latest models are more "robust" and have a "mean global surface temperature within 0.2°C of the observations over most of the historical period... However, some CMIP6 models simulate a warming that is either above or below the assessed *very likely* range of observed warming."

Previous assessment reports admitted the difficulties in modelling and the inability to predict future warming. Modellers feed in inevitably incomplete and sometimes erroneous data, where errors multiply exponentially as time goes by, but the main problem is that we still have a poor understanding of the mechanisms of natural climate change, particularly the part that oceans and clouds play, which interact in such a 'chaotic' way that makes them impossible to predict. Only by assuming that

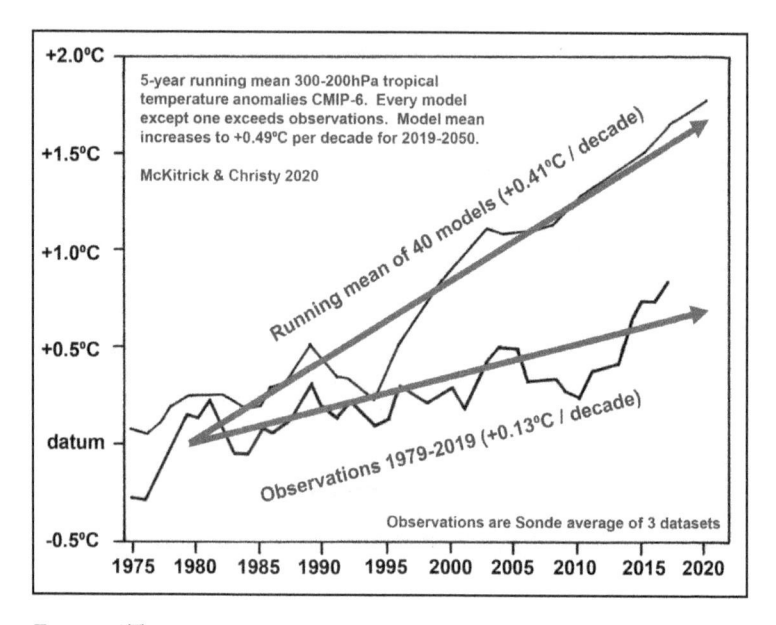

Figure 45 Tropical temperatures: models versus observations since 1979.

the world ecosystem is in balance and fixed, in other words no natural change, can modellers then apply the effect they think human emissions cause, albeit in a very broad-brush way since the possible transient climate response has a very wide range.

Modellers must be eco-pessimists to obtain funding. Imagine if a bunch of academics seeks £5 million to develop a complex computer model. If they go along and say, 'Please give us an enormous sum of money so we can prove that there is not a problem', they will be unlikely to succeed. If instead they say, 'We think there is an existential threat to humanity because of climate change. Please give us the dosh', they will have a far better chance of obtaining funds from the taxpayer or foundation.

To me, the final proof, if proof were needed, that the IPCC is an inveterate, unscrupulous climate alarmist organisation, is that it still peddles the RCP8.5 climate model scenario, now called SSP5-8.5, where world emissions would double by 2050. This was dreamed up over 20 years ago when it was predicted that we would start to run out of oil and gas in the 2020s, that is, now. We would then revert to coal power, increasing our generating power from this high-emissions source from 15 gigajoules (GJ) in the year 2000 to 30 GJ in 2040, 45 GJ in 2060 and 70 GJ by 2100. This scenario was completely destroyed by the hydraulic fracturing revolution, such that the USA's transition from coal to gas made it the country with the largest emissions reduction in the first two decades of the 21st century. The use of coal has increased to 23 GJ because of the growth of China but it is not expected to go beyond this. The Chinese (and Indians) will get sick of coal pollution and convert to more oil and gas as the country gets richer, dropping coal use back to the 15–20 GJ level.

Marcel Crok with a blog clintel.org (YouTube *How Biased is the New IPCC Report AR6?*) found in the body of the report: "The likelihood of high emissions scenarios such as RCP8.5 or SSP5-8.5 is considered low in the light of recent developments in the energy sector (*Hausfather and Potters* 2020a, 2020b)." Crok questioned why this important caveat was not included in the Summary for Policymakers, and I wonder why the IPCC has to refer to an academic study when anyone who reads the newspapers knows that we have not abandoned oil and gas for coal. Roger Pilke Jr (son of a prominent meteorologist, also Roger) calculated that for RCP8.5 to transpire we would have to build 32,000 more coal-fired power stations (one per day) until 2100, and even RCP7.5 would require one new station every second day.

RCP8.5 also assumed that the world population would reach 12 billion before the end of the century. By the time Hans Rosling wrote the book *Factfulness* in 2018 he calculated this to be no more than 11 billion, but this figure is now being reviewed downwards by some people as more and more countries reduce their birth rate to below the 2.1 children per woman needed to achieve parity. Of 167 countries I reviewed (I missed out some small island states), 78 had a birth rate of 2.1 or less, with another 40 at 3.0 or less which should be below replacement level before 2030. That leaves less than 50 very poor countries with population growth, most of which are in inter-tropical Africa.

Unlike the UN IPCC, the UNPD (Population Division) has a good track record for making predictions. While it gets pieces of the jigsaw wrong from time to time, the overall global picture up to three decades in advance is pretty good, with about 9.7 billion people expected on the planet in 2050. What happens next depends on whether Africa gets its act in order, but since the continent currently accounts for only 3% of emissions, carbon dioxide demonisers should exclude it from their reckoning during their lifetimes, because in the meantime economic growth is best for poor people and their environment.

SSP5-8.5, which also assumes a transient climate response of 3°C rather than a more likely figure around 1.5°C, should not have been included in the Report because it is an impossible scenario. It is fodder for the climate alarmists who will use it to claim that we are heading for the 5.7°C rise in temperatures by 2100 that this scenario might have caused. An autocratic bureaucracy like the IPCC, with no facility for democratic public scrutiny, has no capacity to admit that, in RCP8.5 and SSP5-8.5, it got it hopelessly, pathetically and dangerously wrong.

There are three scenarios where emissions reduce, in SSP2-4.5 from about 2050 and in SSP1-2.6 and SSP1-1.9 immediately. There is no chance that the world will quickly stop using fossil fuels, because of the prohibitive cost and the impracticality of such a radical change in infrastructure, even if current alternative energy sources worked effectively, which I contend they do not. The IPCC seems to believe that there might be a magic wand carbon capture solution, but that is just wishful thinking. We cannot predict whether speculative new technology will become practical, nor the timescale it would take to test, develop and roll out on a global scale. So the last two scenarios should not be considered.

I suspect that SSP2-4.5 would only work if the poorest 7 billion people in the world up to 2050 (less than 30 years away) forego the progress towards the wealth that the richest 1 billion currently enjoy, because anything the latter does to reduce emissions will be dwarfed by the emissions of India, China and hopefully Africa, the last continent to banish extreme poverty. In other words, SSP2-4.5 would be morally wrong for rich countries to impose on poor ones.

I would argue that the fifth scenario considered, SSP3-7.0, where emissions would double by 2100, is the only realistic one, although, like the peaking of population growth and then reduction, and the move out of coal, emissions could well have peaked before 2100 and started to decline. The main reason that UK emissions have dropped since 1990 is the transition from coal to gas (much of the rest is merely transferring our industry and therefore emissions to China), and if the West wishes to promote lower emissions in poorer countries, the best thing it could do is make hydraulic fracturing technology available worldwide. The use of natural gas would provide better returns on investment than renewables, but that would require politicians to publish figures on 'green' investment policies, which they dare not do just now for fear of extreme embarrassment. And over the last ten years, Western politicians have become so frightened of climate extremists like Extinction Rebellion that they dare not even suggest gas as a transition to a lower-carbon energy source in the future, like affordable nuclear.

Once we have defined a viable scenario, we can debate what the accurate figure for the transient climate response should be to apply to the increased emissions, and I would argue that it is no more than 1.5°C. So by 2050 I might put the anthropogenic element at less than 1°C, to be added to a natural element of about 1°C, and I would celebrate that we are back to the benign climate we last experienced during the Medieval Warm Period – about two degrees warmer than the Little Ice Age. There might be another 0.5°C–1°C of warming before the end of the century (the IPCC's best estimate by 2100 for SSP3-7.0 is 3.6°C against mine of about 2.5°C), and I would argue that humans are perfectly capable of adapting to such temperatures with current knowledge and resourcefulness. But I would stop before I came to this conclusion, because every other human being who has tried to predict the future has been wrong, even if climate 'scientists' claim to be the modern Nostradamus. Futurology is not a branch of science, but the IPCC is happy to promote it.

8. WHAT IS THE CURRENT OFFICIAL VERSION OF SEA LEVEL RISE AND OCEAN ACIDIFICATION?

The global mean sea level increase was 200 mm between 1901 and 2018 (Para. A.1.7), but the rate of warming varies such that it was 190 mm/century between 1971 and 2006, and 370 mm/century between 2006 and 2018. The latter is presumably meant to cause alarm about sea level rise taking a hike, but *IPCC AR5* was honest enough to point out that a short-term acceleration to 320 mm/century also happened between 1920 and 1950. It seems that our current level of knowledge cannot explain such variations because of a lack of data on ocean heat transfer mechanisms and the part that varying cloud cover plays in imposing wind stress on the oceans.

Figure 46 shows the UK average mean sea level rise from 1900 to 2008, as calculated by *Woodworth et al.* 2009, with readings at Aberdeen, North Shields, Sheerness, Newlyn and Liverpool. I use this to illustrate how one can apply a trend line to suit the occasion. The overall trend is 140 mm per century but within this it is possible to track 30-year trends which go up by more, go down or stay the same, such is the variation in natural climate mechanisms. Such large variations in observations of several aspects of the climate convince many scientists that proposing the concentration of a single greenhouse gas as the dominant factor in climate change is just fanciful. The UK rate of sea level rise is probably less than the global average because parts of the country are still rising back from having three kilometres thickness of ice pressing down on them 20,000 years ago.

In Chapter 3 I also noted that land area increases because of sediment washed down rivers, as in Bangladesh from the Ganges and Brahmaputra, and eroded coral is washed up on Pacific island beaches during storms. The IPCC is happy to alarm us that sea levels are rising, but does not mention that

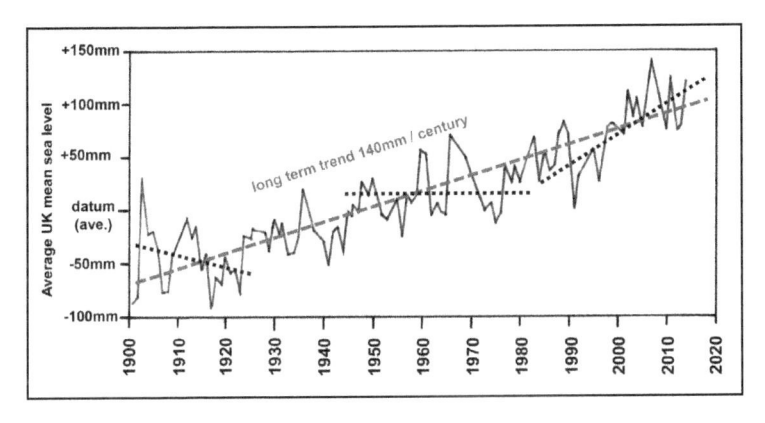

Figure 46 Average UK mean sea level 1900–2008.

Tuvalu, most of the Marshall Islands, Kiribati and the Maldives have actually increased in area during the satellite era when we have been able to observe them from above.

Weather forecasters cannot make accurate predictions of more than a few days, even with the best of modern computer technology and data collection, yet we are expected to believe the IPCC when it projects what will happen in future decades and centuries. "Over the next 2,000 years, global mean sea level will rise by about 2 to 3m if warming is limited to 1.5°C, 2 to 6m if limited to 2°C and 19 to 22m with 5°C of warming, and it will continue to rise over subsequent millennia (*low confidence*)". Why even mention something of which it has "low confidence", and which would not happen for thousands of years? Why mention the 5°C of warming which is at the top end of speculative scenarios, if the intention is not to promote alarm? And could humanity not cope with a 20m rise in sea level over 2,000 years (one metre per century)?

The Report says little about ocean acidification, presumably because it is now seen as an accepted phenomenon, despite *IPCC AR5* only having the evidence of its existence from three observation points (for oceans which cover 70% of the planet's surface) over a period of 20 years, far too little and too short to be statistically significant. It seems logical that more CO_2 in the oceans would lower the alkalinity (potentially bad), but it is just as likely to increase plant life in the sea as on land (potentially good). There is some evidence that phytoplankton is increasing in cool oceans, to the benefit of fish larvae and zooplankton, and all life higher up the food chain. But the IPCC is immune to good news.

9. HOW IS THE SCARE STRATEGY FOR A 1.5°C TEMPERATURE RISE DEVELOPING?

At the time of *IPCC AR5*, the decision was taken to reduce the target warming from 2°C to 1.5°C above pre-industrial times, presumably to try to encourage swift action from governments by winding up the alarm. Report *SR1.5* was issued in 2018, basically saying that if things are bad with 1.5°C of warming, they would be much worse with 2°C, so we should target the lower figure. Common sense would have led to the conclusion that, if the assumptions on the sensitivity of more CO_2 on the atmosphere were correct, it would have been impossible to effect the radical transition to alternative energy sources to stop the 1.5°C level being reached before 2050.

The Report has to admit this, but cannot concede that more than a 1.5°C rise from the Little Ice Age is good (I would, based on the evidence I have seen that the Bronze Age, Roman and medieval periods were warmer than this), so it has to adopt the strategy of accepting warming beyond this level, with carbon capture (it is called CDR in the Report) coming to the rescue to bring it back down. It has to temper alarm with hope. D.1.4: "Anthropogenic CO_2 removal (CDR) has the potential to remove CO_2 from the atmosphere and durably store it in reservoirs (*high confidence*)." This leads to my next point.

10. HOW DOES THE IPCC THINK WE CAN PRACTICALLY REDUCE EMISSIONS?

We have seen how a country like the UK, with underground/undersea voids available from North Sea oil and gas extraction, can consider carbon capture and storage (CCS), but most countries cannot. The sort of quantities of CO_2 requiring to be sequestered would be enormous, and I have pointed out in Chapter 10 that so far only a few schemes like the Quest project in Canada and the Sleipner project in Norway have tested this very expensive technology at scale (Shell uses the captured CO_2 at Quest to force oil out of sands and make the project financially viable, but the Norwegian project relies on large subsidies from this oil-wealthy state and the EU). And yet the IPCC has "high confidence". I reckon it has reached such a high level of desperation that it has to rely on "the potential" of CDR/CCS to sort humanity's problems. In a Report chocked full of speculation, this is the ultimate speculative venture.

The Report is all about scary scenario exploration, and it says little about the impracticalities and costs of strategies aimed at reaching net-zero emissions that I have highlighted in Chapter 10. In section D.2.1 it mentions the temporary drop in emissions due to the restrictions caused by the COVID-19 pandemic, although it says: "global and regional climate responses to this temporary forcing are, however, undetectable above natural variability (*medium confidence*)". It stops short of saying that the radical measures that most activist organisations like 350.org call for (keep all fossil fuels in the ground) would cause a permanent lockdown of the world economy.

We saw in Chapter 10 how the fossil fuel energy used to make, install, maintain and decommission renewables, and to step in as backup when they fail due to their intermittency, leads to a poor return in emissions reduction. But surely things can't go wrong if we plant lots of trees? When asked by interviewer Stuart McNish in his YouTube series *Conversations That Matter* about whether the billions of trees in Canada soak up any extra Canadian carbon emissions that humans produce, Greg Flato replied that trees over their lifetime have a very minor sequestering effect. "Photosynthesis is... taking carbon dioxide out of the atmosphere and growing a tree. But when that tree dies – either it falls over and decomposes or it burns in a forest fire – all of that carbon that it took up goes back into the atmosphere... It's basically carbon neutral... It's not quite in balance. It does take up a little because some of the carbon, a small fraction, is stored in the soil." But this small saving will probably be negated when the carbon-based energy used to plant the trees and maintain the landscape is considered. So planting lots of trees is only a short-term way of reducing emissions whose efficacy will take a few decades of growth to establish, and be lost in a few generations when the trees die.

But, as a 2022 study by Lawrence, Coe, Walker, Verchot and Vandecar, *The Unseen Effects of Deforestation: Biophysical Effects on Climate*, pointed out, it is not as simple as planting more trees to extract CO_2 from the atmosphere, because of the associated biophysical responses to varying amounts of tree cover. Deforestation and reforestation can change albedo (the landscape's surface reflection), evapotranspiration (the movement of surface water to the atmosphere producing latent heat) and canopy roughness (also affecting how heat and water vapour is drawn away from the surface). "Locally at all latitudes, forest biophysical impacts far outweigh CO_2 effects, promoting local climate stability by reducing extreme temperatures in all seasons and times of day. The importance of forests for both global climate change mitigation and local adaptation by human and non-human species is not adequately captured by current carbon-centric metrics, particularly in the context of future climate warming... Beyond 50°N large-scale deforestation leads to a net global cooling due to the dominance of biophysical processes."

So above 50°N reforestation has a net warming effect, which includes Canada, northern Europe (including the UK) and Russia, areas where forest cover has increased over the last half century. This is not an argument to reverse the reforestation process, but an indication of how complex, interrelated and poorly understood natural systems are. I am also sceptical about any such study, no matter how well formulated, because it relies on over one hundred other studies for data and I am sure that the authors did not go back and check how reliable they were, or the studies these studies relied on, and so on. This is in an area of academic research where peer review is the main or only method of quality control, and this is often flawed.

Incidentally, I could point out to Flato and McNish that their country would not have grown and prospered in the way it has without global warming. Even now the Canadian population is stretched along its border with the USA, craving warmth. The Great Lakes and the St Lawrence would not be pleasant places to live if we still experienced Little Ice Age conditions.

Discussion: How many of the 40,000 academic papers on climate change published each year, like the one above on the biophysical effects of deforestation/reforestation, do you think can trace all their sources back to verifiable data, or do they just piggy-back on sources which themselves piggy-back on others? In other words, how much climate change research has a pedigree of solid science, and how much is, in effect, faith-based?

CONCLUSION

The IPCC's mealy mouthed language epitomises the fragility of its narrative and redefines the meaning of 'science' to suit its agenda. Supposing we were to use it in our everyday lives. Interviewee to prospective employer: 'I have *high confidence* I will make it to the job interview'. Clinician to patient: 'I have me*dium confidence* your medical procedure will be successful'. Taxi driver: 'I have *low confidence* that my taxi will start in the morning and I will get you to the airport on time'. I used to have a car like that in the 1970s, and if electric car charging is rationed to prevent grid overload at peak times, this scenario is again possible. I say this with *high confidence* as the UK government is already considering this. Employer to employees: 'It is *very likely* I will be able to pay you next month'. Our lives are not frivolous games. We need more certainty than the IPCC's speculation if we are to drastically change our lifestyles as it wishes. And the solutions it proposes are just as uncertain. There is a 30% chance that the local wind farm will provide useful energy next Monday. This political game is not science.

The IPCC must stop causing alarm about "low-likelihood outcomes, such as ice sheet collapse, abrupt ocean circulation changes", etc. Activists just use them to justify radical action as if they were inevitable. If we were to worry about every possible eventuality, no matter how unlikely, we would not get out of bed in the morning. It should come back in seven years' time with more empirical science and certainty in its next report and, in the meantime, instead of running as far away as it can from critics, address the concerns of climate sceptic groups which have gathered together people who are expert enough to competently scrutinise all aspects of climate research and the energy policies proposed. It is difficult not to conclude that an IPCC assessment report is more political prospectus than scientific treatise, and as an organisation it is beyond redemption.

The UN is in the surreal position of not being able to discriminate between democracies and autocracies, and it cannot admit that its strategy on climate change, by default, has geopolitical consequences that harm the former. Since wind turbines are not blown at an efficient speed most of the time, we still require fossil fuels most of the time, so Germany and Eastern European democracies have relied on Putin's gas to keep their economies working. Substitutes for oil and gas in transportation are some way off (and petrochemicals are needed for essential plastics and other man-made materials), so for the foreseeable future the West will rely on supplies from countries like Saudi Arabia, which has one of the worst civil rights records on the planet. Under the current dumb ideology, hydraulic fracturing in most democracies is still unacceptable, despite the experience of thousands of wells in America which demonstrate that it is no more environmentally damaging than similar industrial activities, including the industrialisation of our countryside with wind 'farms'. Are we also now to rely on the Afghan Taliban for supplies of lithium (as well as opium)? Either Russia or China will supply it with the technology, the former not missing the opportunity to cause mischief in Syria when its support for an oppressive regime resulted in millions of Islamic refugees fleeing to Europe, and the latter keen to increase its current control of 80% of the world's lithium reserves. The UN has, in effect, favoured autocracies over democracies.

In one of his conversation podcasts, Australian politician John Anderson interviewed Scottish historian Niall Ferguson (YouTube *The Utopian Myth of Equality*). They quickly concluded that the pursuit of "equality of outcome" instead of "equality of opportunity", by adopting socialist policies, will emerge from time to time in Western societies because young people who vote today for Alexandria Ocasio-Cortez in America or voted for Jeremy Corbyn in the UK are too young to have experienced the economic chaos in 1970s Britain the last time this was tried. Ferguson, senior fellow at Stanford and Harvard Universities, went on to highlight the three main threats facing Western democracy today. Radical Islam is "explicitly hostile to values of individual freedom, equality of the sexes, [and] the things which have come to be central to our civilisation". Next comes the economic power of China, although he questions how sustainable a one-party state can be in the longer term. Speaking before the invasion of Ukraine he did not mention Russia but, as the author of *The Ascent of Money*, he no doubt reckons that its inferior economic power limits its threat, especially if the West stops purchasing its oil and gas.

But by far the biggest danger is internal. "The most important challenge to Western civilisation [is] the challenge from within: our own self-hatred, or at least self-doubt; our refusal to teach serious history to our kids; our refusal to assign the Western canon in universities." Yet he is optimistic. "None of this really should trouble us too much because, at least on paper, our system is superior. It's superior in every conceivable way. It offers far more opportunity for human self-fulfilment. It offers greater opportunities for gifted people to innovate, to be original. Without free speech how can you really have sustained intellectual advance? That's a fundamental question that I don't think the Islamists or the Chinese have a good answer to. But if we decide that we're the problem, and that all the world's problems from the Middle East to the Far East originate in the wickedness of the West, there's a risk that we just hoist the white flag... It's our decision whether Western civilisation goes down the tubes or not."

Carbon is the basic building block of life. In demonising it, Western intellectuals, from the comfort of their cloistered existence, with calls of "climate change denial" to shut down free speech and intelligent, informed debate, are at the centre of this Western cancer. The economic reality of failed energy policies will hopefully restore common sense, but it sure is frustrating waiting for this to happen.

As early as 1852, Glasgow newspaper editor Charles MacKay in his book *Extraordinary Popular Delusions and the Madness of Crowds* identified a fatalistic herd instinct in various human endeavors from the Dutch Tulip Inflation (1635–36) to the French Mississippi Scheme at the hands of Scottish conman John Law (1719–20) to the British South Sea Bubble (1720). In each, people followed a seemingly certain venture to illogical ends and widespread bankruptcy. As MacKay put it, "knavery gathered a rich harvest from cupidity". He concluded, "Men, it has been well said, think in herds; it will be seen that they go mad in herds, while they only recover their senses slowly, and one by one."

Humanity is sometimes not good at learning lessons but I hope you, the individual reader, have been convinced by my narrative that serious questions need to be asked of our politicians.

POSTSCRIPT

As this book was in the process of publication, in September 2022, the UK found itself with a new prime minister and then a new monarch. The weekly conversations would be fascinating if we could listen in, since any prime minister must strive to reduce energy costs to 'save the economy' of the UK in the next two years before a general election, and the latter is a prominent convert to the 'save the planet' cause. Sourcing more domestic oil and gas is a key element in the former, since the government cannot continue to print money and indebt the nation with subsidised energy costs; but fossil fuels are anathema to the well-meaning but naïve green movement.

From the 1960s, Ladybird Books published Learnabout educational books on various subjects which were popular in schools and very much loved. They had stories with stereotypical models of white British family life, featuring housewife Mum, breadwinner Dad, and the innocent Peter and Jane. It was 2014 before Ladybird abandoned gender stereotyping and began to make fun of itself with comic 'informative' books for adults, including The Hangover, Zombie Apocalypse, Dating, The Toilet Book and The Hipster. This success led to the release of serious science and history books explaining subjects in terms that laypersons, including young people, could understand. The first of these, in 2017, was co-authored by the Prince of Wales and called Climate Change. It is the complete opposite of my book and a prime example of the alarmist narrative for anyone who wishes to know it.

Like his late mother, King Charles III is a forthright, caring and special person, dedicated to public life, and I am as loyal a subject as any. I have only spoken to him once when he came to visit my place of employment in the early 1980s but, if I met him now, I would point out the following observations from the 70 years of his mother's reign.

- From 1952 to the late 1970s there was global cooling and, with the 'hiatus' from about 1998 to 2014, a global warming trend was only observed for less than half of her reign. We do not understand climate change well enough to predict impending doom.
- Since 1952, winters have become two weeks shorter on average, which is marvellous because cold winters are almost twenty times more deadly than warm summers. The modest global warming trend is to be welcomed, whatever its causes turn out to be.
- Satellites tell us that the world's leaf cover has increased in the order of 10%, largely due to more CO_2 plant food in the air. This is wonderful for the environment, especially in semi-arid zones. The modest increase in this life-enhancing gas is to be welcomed, not demonised.
- Due to the human resourcefulness derived from fossil fuel power, deaths from extreme weather events have dropped by around 80%, despite the world population tripling. We don't make a safe climate more dangerous, we make a dangerous climate more safe.
- Cheap fossil fuels are key for economic growth in developing countries and, at a GDP per person per year of around $5,000, people can afford to look after themselves and their environment fairly well. But in the Commonwealth, of which the King is head, 26 of the 56 countries and dominions fall below this, with around two billion people suffering from unacceptable and avoidable poverty.

- Renewables provide 'renewable' energy, but not 'low-emissions' energy because of the dependence on fossil fuels for their existence and to counter their inefficiency. As I have pointed out, green ideology causes more harm to the environment than is alleged for fossil fuels and is a threat to the economic health of democratic countries, while emboldening autocracies.

Sir, please read my book.

Discussion: On the face of it, monarchy should have no place in a modern democracy. It is elitist, hopelessly romantic and takes solace from an unscientific, supernatural Christian tradition, whose morality only makes sense in the modern world if we ignore the homophobic, misogynist, antisemitic, racist, pro-slavery and sin/guilt/cruelty bits of Iron Age scripture. Yet, while the concept of the UK Crown is absurd in theory, it works in practice. It exudes continuity, national unity and dedication to duty, and pays for itself handsomely in tourist income alone. Best of all, it takes ultimate power away from any unscrupulous politician who may come along. How do we reconcile the irrational/romantic and the rational/scientific aspects of society? Do we keep an open mind and focus on actualities and outcomes, or is it not that simple?